Making Sense of

Life

Explaining the Relationship between DNA & Telepathy

Richard Alabone

Revised Edition 2021

Author. Richard Alabone

Essex UK

ISBN: see rear cover

https://makingsenseoflife.uk

Contents

1. An Introduction

Life, as they say, is a mystery. We all wonder about things we don't understand, and assume somebody else will know the answer.

When I was about four years old something happened that I didn't understand. My mother suddenly said to me *think of a number and I'll tell you what it is,* which she did several times. Even at that age, I knew it was very strange, things like that just don't happen, so I asked my mother how and why it worked, but she had no answer. The next day I asked her again and I got the same reply, so I assumed that somebody else knew the answer. Over the years I realised to my surprise, nobody really knows how telepathy works. Unfortunately, I am one of these people who cannot accept that a problem has no answer; if the problem exists, then there must be some way it can be resolved.

Life Seems Mysterious

We are told that science can answer most questions, but scientists regard the evidence of telepathy as fake news, in the same way as religion and science have little common ground. The majority of people accept that the paranormal is fiction; and many believe God made the world so beautiful, while some

1

of us find religious belief too difficult to comprehend. Whatever your belief about these two issues, they surely are mysterious.

So is the miracle of childbirth, and the way a single cell grows into a man or woman. Consider also, the mystery of the mind, an issue which science has failed to understand, labelling it as 'the hard problem'. Inheritance is something we all take for granted, obvious and straightforward unless somebody asks – how; to which there is no answer. Charles Darwin's Origin of Species settled the questions about evolution, but he was never sure if it was mutations, or biological inheritance of family characteristics, that caused species to evolve.

These are some of life's mysteries but the biggest for me is this fundamental question; how does a single cell grow into a blade of grass, or an elephant, or any living thing? This, the scientists say, they have solved by assuming DNA to be an instruction book, which is perhaps very questionable, although they continue to cling to that faulty science.

The purpose of science is to resolve anomalies but scientists have never been able to resolve the problems of life. This is because they can't measure things they cannot detect, or analyse them, so the system fails, anomalies have remained forgotten, overlooked or accepted as a peculiar abnormality. I am a retired engineer who has been able to look at the problems of

1. An Introduction

life in a different way to scientists, by thinking outside the box; anomalies do not exist in engineering.

A eureka moment came suddenly to me, 37 years ago, when I realized that instinct, which is in the mind of all animals, must get there by some process of information transfer from mind to mind, the same as telepathy. This simple but profound idea led me to realise that life is an eternal copying machine, copying all the time from the previous generations. How is it, our DNA, which is mostly scrambled nonsense, provides the information to build our bodies? We are like a house which is built of 30 trillion bricks (cells). I believe DNA cannot possibly be a book of instructions; the information to build the house must come from somewhere else – but where, I was compelled to ask?

Making sense of life has been my goal for the past eighty-five years, and it has involved the personal study of meditation, hypnosis, spiritualism, ghosts, and dowsing; looking for answers, as well as an exploration of many sciences. I followed wherever the evidence led me. Through the course of my studies, I discovered that DNA controls life in a totally different manner to what we have been told. I was quite unprepared to learn that, for the past seventy-seven years Schrödinger, a well-meaning physicist, has led biologists astray, and the obvious solution to the difficulties they face solves many other mysteries and affects other theories. These include the mystery of inheritance, which alters the theory of evolution.

Another theory is that everything we identify as 'paranormal' is actually part of the normal mechanism of life, which telepathically transfers brain and mind information from parents to offspring during the process of growth. When my mother casually said, *'Think of a number and I'll tell you what it is'* I asked *'How could that happen'?* She had no answer, and it was never spoken of again. That started me on a lifelong search for the truth. How does telepathy work and how does it relate to the other mysteries known collectively as the paranormal? And how does that relate to the mystery of life? If taken separately, the individual parts of my proposed theory may appear questionable, but when taken as a whole, the evidence and science of each radical element fits together to form a single hypothesis. If you are unable to accept the paranormal as a possibility, will your beliefs be altered by an explanation of why and how?

By 1984, I had decided that understanding biology was the key, and I lectured to the Society for Psychical Research about a 'rationale linking biology with, special psychic abilities' (psi). Although this was received with interest, I had little evidence to back up my proposal at that time. The evidence only came very slowly, by reading books, personally studying meditation, hypnosis and spiritualism, and doing twenty hours of broadcasting on the topic Paranormal is Normal. (You can listen to them all; links are in the appendix).

Scientists say our genes are the blueprint for life and a book of instructions that ***directly*** control our growth and behaviour. However, although genes ***appear*** to be the blueprint, there is evidence to show that they cannot work that way. This is biology's great conundrum – my evidence gives a new solution to the problems of biology, even though it goes against scientific 'evidences' that have accumulated over the years.

The Royal Institution

I spent many years trying to understand the relationship between instinct and biology realising this was the link that was missing. In 2018, I attended a conference at The Royal Institution in London about contemporary biology, which was sponsored by the European Research Council, and was an investigation into how to fill a huge gap in the science of biology. The aim was to **'rethink the philosophy of biology'**, where the seventy-seven-year-old belief that DNA is a book of instructions was severely questioned, but no alternative suggestions were made. All they did was to go over the evidence showing that the philosophy of biology is unsound. Although none of the lecturers dared to say that genes are not a book, it was clearly in the minds of the lecturers. I was in complete agreement when Denis Walsh, a Professor of Philosophy at the University of Toronto, said, **'Biology is in a mess, so we need a new core theory'.**

My idea from 1984 fitted the bill, the evidence I had accumulated over the years all supported a new theory. If DNA is not a book of instructions, then what could it be? We all know the answer – even children know it. We are copies of our parents; the only question is **how** does it work? I am convinced that DNA is no more than our personal identity code, which again is public knowledge. I am now proposing **'a new core theory'**, showing that genes do not control life **directly** as if it were a book, they are only the key to opening a gate through which the growth information passes. I will show that genes and DNA enable each new generation to be a replica of the last, where the information flows **directly** from the members of the previous generation, and not from DNA.

This revolutionary idea differs from the 77-year-old dogma which is a at the very heart of modern biologists thinking. I make no apology for this because it is based on solid evidence. Biologists need a new core theory, even after 77 years; and it offers a solution to solve many other unanswered questions.

The Unsolved Problems of Biology

So, what are the unanswered questions? What is the mechanism of inheritance; or the mechanism during the growth process that tells a cell how to grow into a different type of cell (differentiation)? How does inheritance information become part of a child's brain before birth? How does inheritance guide evolution of species? And why does telepathy occasionally occur?

It is only by proposing that DNA enables reproduction, rather than it being a book of instructions to build life, that all these questions may be answered. The next question to be answered then, is how and when did the theory of biology go so wrong?

In 1928 a transferring principle was discovered, showing that features are transferred by physically moving DNA from one bacterium to another, but I have discovered that science now shows life may be transferred by the presence of a matching DNA, rather than its physical transfer, which proves the fundamental theory of biology to be incorrect.

In 1981, English author and biologist Rupert Sheldrake wrote about a solution when he suggested in his book *A New Science of Life* that morphic resonance or fields, between members of *any* species, accounted for their growth and form. This concept had been studied since 1910 without success.[1.1] Before publication, the *New Scientist* said of Sheldrake's ideas: '*Of course, within the context of modern science, such an idea is completely scatty.*'

Sheldrake's ideas were controversial because most scientists at the time believed all life grows according to plans in DNA. The editor of the science publication *Nature* referred to Sheldrake's book as one *'for burning'* and subsequently said during a BBC interview: *'Sheldrake is putting forward magic instead of science, and that can be condemned in exactly the language that the Pope used to condemn Galileo, and for the same reason. It is heresy.'* This book contains unorthodox opinions, contrary to accepted opinion or belief. It addresses those same problems that Sheldrake tried to solve, although I believe my solution resolves all the problems.

For over a hundred years, there has been the suggestion that the development of structures and organs during animal growth are caused by some invisible field. But how this may work has never been adequately explained. My findings provide the answers by presenting the evidence that has previously been lacking. It addresses those same unresolved problems and suggests that genes and DNA control life ***indirectly*** by morphic fields rather than ***directly***, offering a mechanism based on the evidence and fresh ideas. This completely new way of looking at biology relies on the obvious fact that we are ***direct*** replicas, or copies, of our parents and their families, where the information flows by morphic fields. I believe we reproduce the structures, organs and brain, directly from our two parents, or their are wider families, rather than from our DNA which serves an entirely different purpose to what we are led to believe.

1. An Introduction

In 1944, the physicist Erwin Schrödinger, who was a Nobel Laureate for physics, wrote a book entitled *What is Life?* in which he said: '*It is these chromosomes . . . that contain, in some kind of code script, the entire pattern of the individual's future development and of its functioning in the mature state ... they are the law – code and executive power – or, to use another simile, they are the architect's plan and builder's craft – in one.*'[1,2] He used the word chromosomes, as DNA was not known at the time, saying they were both the architect and builder of life.

He said this because during tests a transferring principle had been shown to pass detailed information into bacteria by transferring some genetic material. This was in fact an incorrect interpretation of the test evidence. The hypothesis has been accepted as the theory of genes ever since, mainly because no other solution has been available. Biologists are aware that his idea of DNA being a book of instructions is wrong, but they are left without a solution. I am now suggesting a solution, and while I know it is one that biologists will find hard to accept, it nevertheless answers many questions.

To this day, we are still following Schrodinger's ideas, inferring that genes *directly* control life, form and behaviour. We say our genes are the book of instructions, but I am quite sure that's wrong. Our unique DNA code is only two per cent of the total, but it is no more than a key enabling us to copy parental features, while the other ninety-eight per cent of our DNA is no more than a record of the changes to characteristics in our species, which have been carried forward since the beginning of time. That's why we share ninety-eight per cent of our DNA with monkeys. As I have said, DNA works in an entirely different way to what we have been told. It's incredible to realise that a Nobel physicist misled the world for 77 years with his book *What is Life*, in which he said what some people were thinking at the time.

By turning the theory of genetics on its head, a new understanding for inheritance becomes available. We directly inherit information to create bodily form. Whereas previously, inheritance was mysterious, having no explanatory mechanism, it must now be seen as part of a direct copying or replicating mechanism from our parents.

My new core theory, which stems from a new perception of DNA and the full meaning of inheritance, initiates a better understanding of evolution. Amazingly, the large gaps in the understanding of biology are filled, also, psychic experience and the paranormal become normal and legitimate. Detailed explanations for all these claims will be found in later chapters.

1. An Introduction

The Scientific Method

At this point, it's important to explain how science generally works. Firstly, someone identifies a problem, leading to a tentative hypothesis then to experimentation, which may or may not solve the problem. If it appears to solve the problem, then publication leads to others repeating the experiment to test and support the new hypothesis. In this way, individual scientists work to produce tentative explanations or hypotheses, but only within a larger body of science that has been accepted and thoroughly tested in the form of a theory. This is called normal science. Sometimes, however, the theory is wrong. Thomas Kuhn, who was an American philosopher of science, compared normal science to revolutionary science, writing about how an incorrect theory gets changed[1.3] and detailing the precise mechanism.

Any recognised theory (paradigm) can get into difficulty when unsolved problems or anomalies occur. Eventually this may lead to a crisis, during which time another theory is proposed that fits the evidence. The theory in crisis, is therefore gradually replaced by the new theory because it can be seen to fit the facts while avoiding the anomalies.

An example of this is the theory that South America was once joined to Africa. This idea was first noted in 1596, but it was not until 1912 that Alfred Wegener published the idea that the continents had drifted apart.

11

Scientists were not convinced, and Wegener's theory was rejected until 1944 when the manner in which continents drift was pinpointed; although this was still not enough to solve the crisis. Finally, in 1967, it was discovered that mid-oceanic rifts cause the drift of continents and the new theory was at last accepted, which took 350 years for the obvious to be acknowledged by scientists.

Proof was needed that it was more than chance that the outline of Africa fitted America. The idea of continental drift needed a detailed mechanism in order that the principle was acceptable. In biology, the mechanism proposed 93 years ago for the transferring principle is now rejected as can be seen by the European Research Council's searching to *'rethink the philosophy of biology'*. My new core theory offers an alternative mechanism which is supported by much evidence. I will also prove that the mechanism proposed and accepted by everyone, in which DNA is known as a book of instructions, could not be correct. This new fundamental principle and detailed mechanism of biology will be supported by scientific proof.

The Search for Solutions

In my study of telepathy, I realised that the carrier of inheritance and telepathic information is the same as the carrier of the instructions to grow a child in the womb. To make the correct connection the carrier relies on the similarity of the parents' identity codes with the child's code (DNA), like the way a personal telephone number is needed to make a telephone call. The fetus listens to all the DNAs in the world, like a phone, but only connects to the family's DNA which is 50% of its own. This is a complete upheaval for the science of DNA and genetics. In the process, it gives an understanding to the growth of brain and mind, of what we call biological inheritance, and of much more besides.

The realisation of the enormity of my discoveries was hard to bear; I felt I had to write a book to give the detailed arguments and evidence. As a person with a dislike of writing and minimal ability in this discipline, I had a challenge. My first edition was self-published, as no publisher was likely to even consider it. And writing it proved to be a difficult, time-consuming, and frustrating experience. Not only that, I had also hoped the bad writing would be overlooked as the power of my argument, along with the overwhelming evidence, took hold. How wrong I was! The scientists condemned it outright and the general public were bemused. Several book reviewers looked only at the discussion chapter, concluding that it was all fiction. I have had to work long and hard to achieve recognition.

This edition is much improved, updated and properly edited, and it will achieve its purpose in literally *Making Sense of Life*. Over many years, I have looked at the related science, finding that there has always been a feeling that life proceeds by some field (morphogenetic field), but with no accepted suggestion of its origin. Only by looking at the rejected science of telepathy could the origin of that carrier be found. This integrated well with the sciences of DNA, human growth, biology and consciousness, where the mechanism of inheritance leads to the overall explanation of an aspect of evolution that Darwin was unable to solve. By understanding inheritance, I was able to solve what has been called 'The World's Greatest Mystery', which is incorrectly called reincarnation. In fact, it's only inheritance from someone other than the parents.

I spent my life as a development engineer, not as a scientist, but science is only the proper development of ideas. My work achieved several patents involving teleprinters, radar, film projectors and television equipment, where new directions in thinking achieved success. It is unfortunate that scientists often follow dogmatic ideology and conform through fear. For example, even talking about telepathy is taboo.

I believe I have shown that the DNA code is not a book of instructions; it is a unique personal code – one for every person. All life has its own personal code, where ninety-nine per cent of it is the family and species code, which goes back to the beginning of life. All life relies on an unknown quantum carrier of information, which I have deduced from unaccepted science.

1. An Introduction

Darwin searched for a solution to the evolution of species and, although he came up with some accepted ideas, he failed to find an answer to the question of how changes due to environment or desires are transmitted from parent to child, which was the question raised in 1896 by his grandfather, Erasmus Darwin. All life has continually evolved from the early species. For instance, over thousands of years rodents gradually transformed into cows due to a change in environment and diet, making them larger as well as having a different lifestyle and new types of predators. All the time, these changes were propelled by unconscious desires for improvement. Similarly, monkeys came down from the trees and walked about, consumed a different diet, learned how to use tools and turned chatter into speech.

But scientists rejected the idea that needs and desires propelled inherited changes, thinking that DNA was fixed for each species, because all evolution had to be the result of chance mutations. This idea was written into science in 1942 by what was called the Modern Synthesis. Nowadays, the Modern Synthesis is questioned[1.4] so we are still wondering how inheritance and evolution work, how desires are transmitted from parent to child and how the giraffe evolved its long neck. I believe it was the parents desire for a longer neck that changed the neck of the young giraffe, not a random mutation. Its DNA ensured a connecting link was made with its parents, enabling its form to be copied from its elders, but their continual desire ensured that the young giraffe developed a slightly longer neck. As this happened to each generation, it can be seen how the giraffe species developed a longer neck.

The solution to the enigma of life is solved if we realise that all life proceeds by **direct** reproduction of the parents where the information flow is over the ether. This idea is not accepted by science because it has been shown that telepathy is not carried by any known carrier of information. Having looked at quantum biology and the evidence, the details of a quantum carrier have emerged.

In my search for an alternative explanation, I concluded that the unique DNA of every child ensures a connecting link is made with its parents via morphic fields and the quantum carrier. This leads to inheritance changes in the fetus due to the parents changed characteristics, unconscious needs or desires, which can be seen as the driving force of evolution. Scientists regard all this as a wild idea that goes against their dogma, but the evidence taken as a whole shows a different story.

Although it has never been shown how morphic fields work, my new core theory attempts to reveal how and why they operate by an unknown biological information carrier, which then explains much more. As a mechanical development engineer with the Marconi Company working on radar and TV, among the best electronic engineers in the world, I became used to technical research. I have applied that experience to trance, meditation, hypnosis and all the things that seem unsolvable.

The rest of this book is about resolving any challenges there may be in following my new and unorthodox ideas, giving detailed evidence and representing a giant leap forward, which I believe does explain how life makes sense

Making Sense of Life

2. The Things That Don't Make Sense

There are many things we do not understand in this complicated world, and so we rely on our scientists to explain things to us. In practice, they often say that something is *'not fully understood'*, meaning it's a complete mystery. This chapter is all about things that are *'not fully understood'* but at the same time are evidenced by irrefutable facts which are overlooked or ignored because science is unable to resolve them.

The Problems of Biology

Although microbiologists have discovered the most amazing details regarding the chemistry and structure of cells, when it comes to the way cells split into different types to form the various parts of a structure, there is no indication of a mechanism. It's *'not fully understood'* and is a complete mystery. How is it that every cell in our body, which all have the same DNA, can change into about two hundred different types of cell, all doing different things? For seventy-seven years, we have been told that genes are a book of instructions, but this hypothesis could be in doubt, as no scientific mechanism has been discovered. So, my question is: where is the book?

Geneticists insist that the genetic code in the fertilised embryo cells of a species contain the blueprint for the whole organism, although by way of explanation they can only give vague notions of epigenetic landscapes or biological fields. If genes do not ***directly*** organise growth, could they not do it ***indirectly*** by some unknown method? In 1961, the French biologist and Nobel laureate François Jacob said that very little was known about the regulating processes in embryonic development (transcriptional regulation). This remains the case to this day.

We can all notice our family likenesses, which are clear evidence for heredity. It looks as if the whole blueprint, or book of instructions for growth and form, is in the DNA of every cell. This appears to be a reasonable assumption, but is it a likely scenario? I think not.

It has been shown that our genes, which are personal to each and every one of us, are associated with our inheritance due to changes in parental lifestyle and environment. In fact, the accepted doctrine that each species has its own fixed DNA is wrong. Characteristics acquired by an individual during his lifetime have been shown to alter his children's genomes, but there is no evidence of a mechanism.

Scientists have solved the mystery by saying the evidence is above genetics, and calling it epigenetics, but they have not decided how the change becomes encoded in the genes. The different proposed mechanisms for epigenetics are chemical observations of change, but not mechanisms.

As humans have superior intelligence, it was assumed we would have more genes than any other species. The number of genes measured in a simple thing like rice was thirty-eight thousand, while our gene count was only twenty-five thousand. No one can explain why this is, and as already mentioned, we share all but two per cent of our DNA with chimps.

There is talk of a gene for tallness, or one for a particular disease, but it has been found that genes don't work like that: there is much disagreement as to which genes are 'for' a particular trait. Geneticists have spent billions trying to sort this out, all to no avail and it's still *'not fully understood'*.

And, of course, there is non-coding DNA, known as junk DNA because it appears to do nothing. Surely there has to be an answer somewhere to all these problems, but scientists are confused and have no answers.

Biologist Rupert Sheldrake noted that while genes may be the instructions to build a living organism, there is no conceivable way that they could control nonstandard growth, for example, regeneration, reproduction, and regulation.[2.1] He relied on morphic fields but could not explain HOW they work.

Taking these three in turn, regeneration is the process whereby growth proceeds when part of an organism is damaged; it will restore or replace the damage. For example, a flatworm cut into several pieces will grow into several flatworms. Similarly, if you remove the lens from a newt's eye it will regrow from the iris, but the original growth was formed from the skin and by a different mechanism.

The reproduction of a detached part of an organism is well known. For example, plant cuttings will regrow by forming roots, or a plant grafted onto another will continue to thrive. Similarly, a transplanted human organ can be persuaded to replace one that has been removed.

Regeneration and reproduction both require some system of regulation of growth other than that provided by the genes. There must be *something else* regulating this growth. Biologists have no answer to these problems, and for them to suggest that all growth and form is due to some unknown kind of *self-assembly process* is just not good enough. There is no such thing as self-assembly.

Common to Every Living Cell

All life, from bacteria to plants and mammals, has two vital things in common: the first is DNA, which appears in every living cell, and we have been able to examine DNA in the most minute detail, but scientists are unable to describe the physical and chemical processes as they are *not fully understood.*

The second thing that is common to every type of living cell is microtubules. As the name suggests, these are microscopic tubes. We know what they look like and what they are made of, and we understand that they organise cell splitting, but beyond that we have absolutely no idea how they work or why they are fundamental to every cell appearing in all forms of life. This is a great mystery. Why is it that all types of life, from bacteria to mammals, have these strange structures? They have evolved into several different complicated forms, which show that they are fundamental to every form of life. There will be more on microtubules later.

Instinct

If we believe that the instructions for all growth are in the DNA, then that would explain the mystery of instinctive memory. The fact is that biologists haven't been able to find any mechanism for growth by examining DNA, let alone a mechanism that would copy instinctive information as part of the growth process.

Each animal species has its own innate instinctive behaviour. Cuckoos, when they are young, fly to Africa to spend the winter in the warmer weather, but in the spring, they return to where they were born and find another bird's nest in which to lay their eggs. How and why, they do this, we just do not understand. How all this complex information could possibly be arrived at from DNA is a complete mystery: there has to be an answer.

Consciousness

No less mysterious is our brain and mind. We have perception, goal directedness and creative intelligence, and these attributes can be seen throughout the animal kingdom. There is no explanation showing how these could just appear out of the blue. You can find thousands of scientific papers about the various aspects of consciousness on the internet, but they have almost nothing to say about how grey matter grows or works, or how it thinks. One paper that appealed to me said that to understand consciousness there needs to be an added ingredient, which I am sure is true. Whatever could that added ingredient be?

Trance

Psychologists and scientists have spent much time and energy trying to understand the various examples of trance – which is merely accessing the dream state – although some of them deny its very existence. But we all sleep and dream, which is a start at understanding it. Sorting out how and why has eluded the professionals, but I have previously published a book called *Sublimity* containing my own thoughts on this matter. As well as dreaming, there are lucid dreams, which are dreams controlled consciously. Also, we can sleepwalk, which is clearly unconscious. Many people deliberately relax into trance, which we call meditation, or we can have someone else guide us, which is known as hypnosis. Other mysteries related to trance include mystical experience, healing, near death experiences and déjà vu.

Although not the complete list, trance is an enormous subject as well as being difficult to understand. Many wild explanations and beliefs have been brought about by misinterpretations of the evidence, but trance does not have to be thought of as mysterious: I believe there is a straightforward explanation – to be discussed later.

Mind to Mind and the Paranormal

If you believe in mind to mind or telepathy, then it's true to say it changes everything, but if you don't, read on. In science there has always been a taboo against even mentioning the word telepathy. Scientists say that there is no way it could happen, and so it is never discussed. All the evidence must be fraudulent, the result of poor experimentation or just wishful thinking. But could the scientists be wrong?

My childhood introduction to telepathy with my mother gave me a belief in its truthfulness but, as I grew up, I also heard about ghosts, spirits, hypnosis and reincarnation, none of which made any sense and made me as sceptical as the next person. Later, I joined the SPR (the Society for Psychical Research), which examined about twenty different phenomena, including spiritualism.

I've spent much time investigating the possibility of something remaining after we die. My study of spiritualism reveals there is much apparent evidence for something remaining after death, evidence that should be accepted by scientists because tests have shown that sometimes correct information, given by a medium, cannot be rationally explained. When it comes to spiritualism, could there be another channel for the information? We don't necessarily have to invoke the spirits.

Occasionally, another very strange thing happens sometimes when a person is resuscitated; they experience what is known as a near death experience (NDE). A similar strange state of mind can happen at any time, known as a mystical experience. Could there be a relationship between these?

My SPR studies, included sitting up all night waiting for a ghost to appear, buying divining rods and attending lectures and conferences, were all to no avail. I read about reincarnation, precognition, map dowsing and remote healing, which all added to my general confusion, while looking for a common-sense way they could possibly work. Those sceptical beliefs changed extremely slowly, but this book may allow a quicker change in your point of view.

And what about reincarnation? Millions of people around the world believe in rebirth, or the transmigration of souls, which forms part of the faith of many religions. In the West, reincarnation is generally regarded as a ridiculous superstition, which used to be my own belief. So why have I changed my mind? In a word: *evidence.*

In the book *Where Reincarnation and Biology Intersect*, Ian Stevenson gives detailed evidence of sixty-five cases, and he collected more than two thousand more. Typically, a child will remember a past life and give a name or location. Often, birthmarks or defects indicate the manner of death. This transfer of information to a child, rather like inheritance, has been interpreted as being reborn or reincarnated. The evidence can't be contested, as it's been proved beyond any reasonable doubt, following the examination of two thousand six hundred cases by psychologists. But what if this evidence has been misinterpreted?

Reincarnation has been called one of the world's greatest mysteries (more on this later), but it is just one of the psychic experiences investigated by the SPR that have no possible explanation and make no sense.

And lastly there are telepathic twins. They sometimes claim to sense when the other is in trouble, or just know what their sibling is thinking about. The idea is accepted by every culture, but no one can offer an explanation. It is *'not fully understood'* and a complete mystery to science.

There appears to be about twenty different phenomena, all labelled as paranormal, all supported by sound and oft-repeated evidence. Could it be that they all have a common root, which I like to call mind to mind or telepathy?

Could There be a Carrier for Telepathy?

Many years ago, telepathy was called mental radio, which was a good description of what it is, although scientists have shown that it is not a radio system using an electromagnetic carrier, like heat and light.

It must be something else, something that scientists don't know about. This leads to the question: what could be the carrier of telepathy and morphic fields? Could it be more to do with quantum science, which is all pretty difficult to make sense of anyway?

The thought that there could be an unknown carrier just doesn't make sense, and science condemns the possibility of there being one, although I will show there is sufficient evidence for *what* the carrier is and the *way* it works.

Theory of Evolution

In 400 BC, Aristotle wrote down his ideas on the subject. Others then added various theories until Erasmus Darwin documented his observations and thoughts, followed by his grandson Charles, who realised the importance of natural selection and suggested two possible means by which mammals could have evolved – random mutations or the inheritance of characteristics altered by use or disuse of a part of the body. Either of these mechanisms, which cause diversity, gradually alter the evolution of a species. But a century after Darwin, the idea of the inheritance of acquired characteristics was completely discredited by a large body of scientists, as it could not be reconciled with a fixed species DNA, leaving only mutations.

The accepted neo-Darwinian idea, or modern synthesis, is that chance mutations coupled with natural selection account for evolution. However, Sir Fred Hoyle, an English astronomer, once compared this process of chance with the likelihood that, *'A tornado sweeping through a junkyard might assemble a Boeing 747 from the materials therein.'*

Today, the science of evolution is incomplete because there is no acceptable mechanism available to explain inheritance and diversity. Even though Erasmus Darwin raised the problem as far back as 1796, it's still *'not fully understood'*.

The beauty of all life must be wondered at, and it appears to have been produced by the hand of an intelligent designer. Surely, the evolution of life must have been guided by something or somebody desiring change. I discuss the likely source of those desires later in this book.

Is There a Solution?

This long list of anomalies appears to be difficult to understand, apparently because science has not been able to measure or understand the anomalies I have listed. Microbiological research has failed to discover the ways in which the genetic code organises and controls growth, although at long last it's been conceded that there must be something more than the genetic code, hence the term epigenetic, meaning above genetics. Consciousness studies have been similarly fraught, with psychology and science left completely baffled when trying to make sense of the mind. Parapsychologists have looked at the twenty inconceivable types of strange evidence and labelled it all as 'paranormal'. Each one is a paradox, meaning something completely absurd, which at the same time could be true. In fact, although they seem absurd, there are real answers to those things that don't make sense.

Worry not. It's my hope that you will find all the anomalies noted in this chapter satisfactorily addressed in the rest of this book.

Making Sense of Life

3. The Wonders of Biology

Biologists have made wonderful advances, for instance, writing out the genome 'book of instructions' and designing a vaccine to beat a worldwide pandemic. But they are aware of unanswered questions and **wonder** why there are so many that are '*not fully understood*' and they **wonder** what could have gone wrong with the scientific method.

I think by now I have made the point that biologists have a fundamental problem in their understanding because genes do not work in the way imagined 77 years ago. I am proposing that DNA works **indirectly** rather than **directly**, where life proceeds by always copying the previous generations. This is achieved using the DNA digital code ensuring that, rather than receiving building information from its DNA, the information received is from the two parental families where DNA checks that the incoming information is **only** from each parent.

A child's DNA is a random mix of the parent's DNA, which will approximately indicate physical similarities, as has been discovered by geneticists and is fairly obvious to everyone. However, I have realised that parental DNA enables the growth of a fetus, by acting as a key to the message, rather than being the actual message of growth. This change in the way we think about genetics, alters nothing in understanding the wonders of biology, except that growth messages are copied **directly** from parents rather than from what is wrongly perceived to be a book of instructions.

Contrary to scientific belief, this is how all life works. For the rest of this chapter, I will give the arguments and evidence to support my case. I understand that these arguments will challenge many biologists who may refuse to consider the evidence presented. On the other hand, some may well see that the new core theory is the way out of the current predicament, which has held back research into biology for decades.

DNA, Genes and Cells

Any biologist working on research must base his or her thoughts and studies on the theory that DNA is like a book of instructions, because any experimental results that do not agree with this theory must be discarded. This unfortunate situation means research efforts have been wasted and anomalies have been ignored, which hides the fact that the fundamental theory is incorrect.

Here is a very broad outline of DNA.

Scientists have examined the DNA molecule in the smallest possible detail and can see that it is a long-coded molecule of four different chemicals, which identify a particular individual in code form. Ninety-eight per cent is common to tribe and species, and only about two per cent is relevant to that individual and his or her family. Moreover, the long length of code is divided up into identifiable and numbered lengths called chromosomes and is then subdivided into genes. We have twenty thousand genes and forty-six chromosomes, whereas yeast might have only six thousand genes and sixteen chromosomes.

The details of the human code have been laboriously transcribed into a 'book' of about three billion letters, which mostly appear meaningless. This is known as the genome. A similar code study has been carried out for various animals, plants, and bacteria. All this science has clearly identified that DNA is the controlling factor for life; it's only the way it controls life that is in doubt. In some ways, there appears to be direct control from DNA, whereas the majority of DNA seems to do nothing. Biologists have become more and more frustrated by this horrible situation. The DNA code looks like a book of instructions because it bears a resemblance to each individual, but at the same time there are reasons why it cannot be such a book.

Moreover, life is made up of about two hundred different types of cells, all containing the same DNA factor, controlling the growth of the heart, skin, and bones, etc. It's all terribly mysterious and biologists are frustrated, although they don't like to admit it.

So, how did the science of biology go so wrong? To find the answer, we must go back to 1928, when a British geneticist called Frederick Griffith found that there was a Transferring Principle in what we now call DNA. Then in 1944, a landmark paper by the Canadian American physician and researcher Oswald Avery and colleagues, demonstrated the Transferring Principle by moving DNA from one species of bacteria to another, which they assumed transferred the inheritance traits across and showed that the instructions to build the next generation were in the DNA.

Although it seemed a reasonable assumption at the time, this fundamentally flawed idea is where the science of biology went wrong; they misinterpreted the evidence. Scientists did not realise that the transfer of DNA allowed the **direct** copying of the traits from the donor to the next generation, rather than transferring a book of instructions. In that same year, Schrodinger answered the question 'what is life?', by saying chromosomes are the architect and builders craft in one. This is what caused the fundamental flaw in biology research, which we still have to this day.

A classic in-depth book, *Microbiology of the Cell,* was published thirty years ago [3.1] and gave details of the cell mechanisms that control all life. When giving information regarding how genes regulate cell growth and division in the human body, the authors said the processes were *'poorly understood'* and the question of what is responsible for cell growth *'largely unanswered'*. Similarly, analysis of the chemistry of the extracellular control of cell division shows that it is not DNA but *'uncertain mechanisms'* that control all life. My quotations have been selected to show that great efforts have been made to find and present the controlling mechanisms of life. In each case, chemical changes were observed while looking for the mechanism, although it was not discovered. This shows that although a certain chemical change may resemble a mechanism, in the end, all life comes down to the uncertain mechanisms of the growth process, for which there was no answer.

Today, scientific writers are very reluctant to infer there are any problems with genetics and biology, although those same mechanisms remain undiscovered. I have in my collection a beautiful book called *Developmental Biology*[3.2], which contains about four thousand colour plates and drawings and cites over five thousand papers detailing the mind-blowing results of thousands of amazing experiments with single and multi-cell animals and organisms. These provide physical and chemical details, but as far as I can see they're all based on the incorrect theory that DNA is like an instruction book.

As a result, the authors of this book note[3.3] that there are six fundamental unanswered questions in biology. I have paraphrased their descriptions and, in each case, they ask the question **how,** which I have highlighted as follows:

The question of differentiation. During a mammal's growth, the single fertilised cell repeatedly splits – known as differentiation – into about two hundred different types of cell to form skin, heart, and lungs, etc. **How** can that be?

The question of morphogenesis is the question of **how** the two hundred different types of cell organise themselves to form the heart, lungs, and brain, etc.

The question of growth is the question of **how** cell division is tightly regulated to produce a copy of the form of its species?

The question of reproduction concerns **how** the very special cells of sperm and egg transmit the instructions to make a new organism, and what are those instructions?

The question of evolution involves the inheritance of particular characteristics belonging to the parents.

How is it that there can be new body shapes, which can lead to a new species?

The question of environmental integration asks **how** development is influenced by the environment. For example, how the giraffe developed a long neck.

Despite an incredible amount of research and wondering about biology, these six questions are still unanswered and *'not fully understood',* although some have supposed explanations that are not mechanisms.

I believe the reason that these problems are unsolved is because researchers are forced to work according to an incorrect paradigm. There is always the suggestion that a little more research into ever more detail will reveal the automatic mechanism that will explain the challenge of making sense of life. Over the years, it has become obvious that there is something wrong in biology, but none of the ideas or experiments have suggested a necessary paradigm change.

The great conundrum is that we have found the DNA code does not identify a particular individual's hair and eye colour, as was once thought. However, it's obvious to us all that features like hair and eye colour are inherited; but inheritance is *'not fully understood'.* In fact, inheritance is not mysterious if it is viewed as being replicated, down through the generations, by direct copying of the two parental families.

Having said that, science has identified parts of an individual's genome that are clearly responsible for inherited disease. My late nephew had a gene variation identified as a nonstandard chromosome number fifteen. It led to him having identifiably different features, drives and personality. This condition is known as Prader-Willi syndrome. People with this disorder can become very overweight because they have an insatiable desire to eat.

Several other syndromes also have a nonstandard chromosome 15, including Angelman syndrome, in which there is intellectual disability causing walking and speech difficulties. Also, in other chromosomes, there is the Autism spectrum disorder, and about forty more inherited 'diseases'. I can personally understand this as I am on the autistic spectrum myself.

I like to think of those who have had such a diagnosis as different tribes of people, who are subject to Darwin's survival of the fittest and natural selection, rather than people with a genetic 'disease'. They result from the way the parental DNAs merge, sometimes producing a 'normal' child, sometimes not, depending on the hazardous mix of the parent's DNAs, and those of previous generations.

The Process of Growth

At this point I should say that DNA is short for Deoxyribo Nucleic Acid which has been replacing Ribo Nucleic Acid (RNA) as it is more efficient.

Millions of years ago, life was composed of single cells with no nucleus (prokaryotes), which gradually evolved into larger multi cells with a nucleus (eukaryotes). The genetic information of these plants and animals were stored in the nucleus, originally RNA and later DNA. However, today some prokaryotes are multi-cells using RNA, and some eukaryotes are single cells with DNA: life tries all possible ways to change in order to evolve into being a better survivor.

Bacteria, slimes, and moulds ruled for 2500 million years. But the big change came with the evolution of a single celled creature, for example a paramecium, which swims in our garden ponds today and has layer upon layer of complexity, exhibiting all the intelligence needed to survive in its environment. A paramecium is quite fascinating; it's only about 0.25mm in length, so it can be seen if you have good eyesight. It is covered by many tiny hairs (cilia), which it beats in order to swim and find its food. If it bumps into a solid object it will beat its cilia the other way to reverse, then turn and swim away. But it has no sense of sight, taste, touch or hearing. It has a mouth, gullet, stomach (vacuole) and anal pore, which amazingly is all in a creature only a quarter of a millimetre long. All single cell creatures seem to work in an organised and purposeful way, but there is no indication of how this could have come about.

Cells eventually started to group together, leading to cell specialisation. Plants are multicellular, which enables them to become very large, with roots to take up water and leaves, where different cells can capture the sun's energy. The thing people don't like to accept is that we are simply a single-celled creature ourselves, which has learned to be very efficient by repeated division of the original cell, which during our lifetime grows into a hundred trillion cells all contributing individually to the function of our body and mind.

Human life begins as a single cell, which divides repeatedly before the cells start to specialise. By week three, our cells form into three layers, which fold into the head, body and the start of the limbs, with a beating heart by week five. From week seven, the embryo is termed a foetus. That process of growth, from a single cell to a mammal, has been no different for the last two hundred million years.

Scientific descriptions of the growth processes have inevitably led to a lack of success. A paper on plant growth notes that the fundamental principles of development and tissue organisation are '*essentially missing*'. A 2013 paper on microtubules[3,4] and the regulation of gene expression says: '*However, the signal transfer production mechanisms involved in this communication are little understood.'* In another paper, on the analysis of nerve fibre growth in the brain, it suggests that the regulation of this process *'is poorly understood',* and later, *'The mechanism of this activity is still mysterious.'*

Biologists have studied stem cells, which appear to be the key to life. They split in two, sometimes into two identical stem cells, sometimes into two specialised cells, and sometimes into one of each but it is not known how or why. This is the process that governs the growth of a fetus, or any living thing. In some complicated way, a stem cell will produce the necessary chemicals to differentiate into the correct type of specialised cell – be it for blood, liver or bone – in the correct place to produce a fetus. But the process of differentiation remains a mystery.

For the first few weeks, human cells will develop in the same way as a mouse or a horse, before they start changing into a primate. Later, they become a particular species of primate, with a larger brain. Only in much later stages do the data from race and family come into play to develop facial features and personality. That building information, or data, must come from family members, where DNA is like a few phone numbers: half the numbers from the father and half from the mother. These codes (phone numbers) are used to check the origin of the data used for the building process.

Much information about the process of growth has been discovered, except that a false idea, brought about by incorrect interpretation of the evidence, has dominated research. My alternative hypothesis, reliant on a quantum information carrier, views DNA growth processes in a new light. If the evidence and science of a quantum carrier, detailed in Chapter Five, is accepted, it may lead to a new way of understanding DNA, and indeed the whole of life.

In this search for a mechanism that creates life from DNA, we must now accept that the breaking of the DNA code has failed to produce answers explaining **how** and **why**, which would make sense of the process of growth.

The Barcode

DNA is a chemical code to identify its owner, rather like the barcode we have on shop items. During growth, it identifies incoming signals for the copying process of the body and brain.

As far as we know, 3.8 billion years ago, bacteria and algae evolved and probably relied on some sort of memory to aid growth and replication, which was stored as a chemical code (RNA). Importantly, the code also had to store information regarding changes in the environment. As different types and species of bacteria evolved, their barcode recorded differences between one type of bacteria and another, in the form of an identity code; which explains why and how every individual has its own DNA code.

The code is made of a long chain of four chemicals and most life relies entirely on the machinery of this barcode. In humans, every living cell contains our own unique chemical identity code that grew from chemicals in the food we eat, and we have a hundred trillion of them in our body.

The machinery of each cell is mind-boggling. Science has discovered all about the chemistry of the barcode, using words such as codon, mRNA, genes, nucleotides, double helix, RNA polymerase, reverse transcriptase, and the genome, all in an effort to find how it works. What they have failed to discover, is what makes cells change, known as differentiation. When cells split, how do they know whether to differentiate, or change, into blood, bone, or muscles? This mechanism remains elusive: all that can be said is that it happens automatically. But as an engineer, I know things don't just happen automatically. It seems to me that the chemistry of cell differentiation is a reproductive or duplication system, which started at the beginning of life with the replication of bacteria. Now, all life uses exactly that same replication process, either for building a blade of grass, a mouse, or a man. And it's a copying mechanism, duplicating all the details of a blade of grass, a mouse, or a man, where each cell either duplicates itself or changes into a different type of cell. The purpose of the barcode is to identify the correct quantum information to allow the cell to differentiate correctly.

The vast majority of the barcode has been labelled as junk, because its full of repetition and apparent errors, but it is not junk at all. In effect, it is the evolutionary history of a branch of the tree of life. That's why our DNA is ninety-eight per cent the same as other primates. It is not junk because it identifies the particular species. Even though it is barely recognisable, having suffered mutation damage over millions of years, it still works as a key. But despite our analysis of the chemistry of DNA, there is still no clue regarding what it is that selects the way cells divide to form the various parts of an organism. I suggest it has to be a duplication mechanism reliant on a quantum carrier, which will be detailed later on in this book.

The sections of the barcode that appear to be useful (coding DNA) have been studied in great detail in the hope that they might be better understood, and junk (non-coding DNA) is being examined by a research project called ENCODE, that cost a total of 400 million dollars. The goal was to find the connection between genes and human traits, bodily features, susceptibility to diseases etc., but they found there was little or no connection. We have certainly moved on with our study of the barcode, even though scientists have had to ignore the awkward problem, such as how a hundred trillion identical DNAs in our body produce two hundred different types of cell during the process of growth. This has caused much difficulty for scientists.

It's been found that one thousandth of an individual's barcode is special to them. This can be isolated and identified by a DNA test, which is used to test for parentage, or individual identification in a court of law, where DNA is left at the scene of a crime.

It was thought that the analysis of a genome would give a personal medical indication of susceptibility to particular illnesses, which might aid treatment. This was thought to be a good idea, but it had little to no foundation in fact. Tim Spector, Professor of Genetic Epidemiology at King's College, London, decided to have his own DNA checked by two companies in the field. He then compared his personalised genetics results. He said in his book[3.5] *Identically Different*, '*Each common disease was controlled not by one gene but by hundreds or even thousands of genes . . . another consequence was that for common diseases these gene tests were pretty useless for prediction . . .*'

This is an example of the fundamental misunderstanding of genetics. We are all fully aware that we inherit temperament, looks and certain abilities directly from our parents, and we are told that this is because of our genes. But that idea is wrong. Our genes or barcode do no more than identify us. We know we inherit features from our parents, but we inherit 50 per cent of our genes from each parent purely as a means of identification.

It is obvious that everyone is a unique member of its species, where detailed information is copied or inherited from the two parents' families. Here the words 'copied' and 'inherited' are synonymous. Inheritance is controlled by DNA, not by directly using information contained in the DNA, but by using information copied from the two parents and their families. The new core theory indicates that DNA ensures that inheritance only occurs between parents and offspring, whether it's a fish or a mammal. This is my new understanding of inheritance, and all life works in this way.

Scientists have been able to create new varieties of crops, animals or bacteria by moving bits of DNA from one species to another. Genetic modification has produced benefits, which are safe, but many countries reject GM because they fear the occurrence of an inadvertent mistake that might not be noticed until it is too late. For example, we know that crops are modified to produce their own insecticide, which then gets transferred to other plants. So this could lead to the uncontrollable decline of insects or fish, which would be disastrous to a delicate ecosystem. Is it worth the risk?

The situation with gene editing is worsening because it is becoming so easy. In China, there has been human gene editing to cure disease, which has prompted a global outcry, as well as an investigation by Chinese health authorities. The unregulated research has been condemned as 'madness' by one hundred and twenty Chinese scientists, who suggested that DNA changes would be passed to future generations, having unforeseen effects on the entire gene pool. I am concerned because I believe the work is being done to the wrong paradigm. A genetically modified barcode does not modify instructions to create a new form or trait, but the modification enables the **direct** copying of form or traits from other species. This produces the same results, but by a totally different mechanism.

Just as our phone system works, life's barcode is transmitted round the world, identifying the sender and the messages. Our mobile phone listens all day for its number, which it identifies to receive the caller's speech. Similarly, a fetus listens all day for its barcode, which it identifies to receive its parents' growth messages. We take our phone system for granted, so why not the barcode system? A hundred years ago, the telephone system was very primitive, so the idea that a small box could communicate with the other side of the world would have been ridiculous. But the evidence is that our barcode is carried round the world, so the idea is not so ludicrous.

If I buy a phone, it will have a barcode on the box that identifies the make of phone. It is not the instructions to build the phone, like the scientific dogma says it is, it's only a code to identify that phone.

I like to think of DNA as a very old book of telephone numbers, where most of the numbers are out of date or junk, but new numbers are constantly being added: one for every new birth.

Epigenetics

Biology has had a basic problem for many years, in that some evidence does not fit the idea of genes controlling growth.

Darwin's evidence of changes to the shape of birds' beaks, caused by the food they ate in the Galapagos, shows us that form and habits can change over time, but this does not accord with the permanence of DNA. The evidence has shown that looks, mannerisms, phobias, etc, are inherited by a child – we all know it - and

Lamarck said it in 1809; but the scientists say it's the DNA instructions that control growth. The result of this longstanding error is that the science of genetics has had to be altered; there needs to be more science to explain the difficulty. And that science says that, above the normal theories of genetics, there must be something else to control growth. That something is epigenetics (above genetics).

To explain how genes control the progress of growth, the hypothesis of epigenetics states that the DNA genes are switched on or off by proteins manufactured by the junk DNA (DNA methyltransferase). This strange idea developed because of the mounting evidence that changes in human environments, thoughts and deeds are passed on through the generations.

Epigenetic theory is the same as saying that DNA has a recording system, which in some way alters the operation of DNA in the following generations until it is gradually removed by blending inheritance. Any reasonable evidence of a mechanism for this hypothesis is lacking.

Jon Baptiste Chevalier de Lamarck was unable to describe how thoughts and deeds are passed down through generations, but he was quite sure he was right. His theory of evolution by inheritance of acquired characteristics was acceptable to some, including

51

Charles Darwin, but many preferred the notion of divine creation. In consequence, Lamarck is generally only remembered for his highly criticised theory. Even in France, he was ridiculed and then forgotten. Eventually he died penniless and blind, and he was buried in an unmarked pauper's grave.

It seems to me that the observed chemical changes attributed to epigenetics are the result of inheritance rather than the mechanism of epigenetics. Surely *some other* mechanism, which organises all growth, is responsible for these chemical changes.

In order to get epigenetics into perspective, we should look now at some of the evidence which has necessitated this modern fundamental change in scientific opinion regarding the idea that DNA is the information source of life.

The evidence often quoted is that when a type of mouse (agouti) has offspring, they alter in size and coat colour depending on how the mother is fed during pregnancy. If well fed, a normal, yellow, fat mouse will give birth to a smaller brown mouse. Later, when her offspring reproduce during a normal feeding regimen, their young will inherit the same change as their parents, being smaller brown mice, and so on for several generations. This rather bizarre effect was noted over a hundred years ago and has puzzled scientists ever since. Another test that showed epigenetic results (trans-generational inheritance) was on chickens whose parents were subjected to irregular lighting, causing them to alter their feeding habits at the time the eggs were laid. It was found that the resulting chickens inherited the altered feeding habits.

Similar results have been noted in humans. Dutch women who were undernourished during the first three months of pregnancy due to the Second World War produced babies who in later life tended to be obese, as were their children. Records show that in Sweden, boys who were undernourished fathered boys that tended to have high blood pressure and heart problems in later life. There are many examples of plants and insects inheriting changes due to environmental shifts.

All these cases are well documented and are examples of trans-generational inheritance, or, to use Lamarck's words, the '*inheritance of acquired characteristics.*' Science has had to come to terms with the fact that this revelation has shattered the central dogma, leaving a rather unfortunate scientific mess, leaving a gap in our understanding of DNA, which has been overcome by the idea of epigenetics.

It is claimed that queen bees are a classic example of epigenetics, where the queen's DNA is the same as her sister workers. How could this be, the scientists ask? It's clearly something different to normal genetics, so they say it must be epigenetic. In reality, workers are just stunted queens, because they are only fed royal jelly for three days before going onto pollen and honey. They can still lay eggs, but they normally leave it to the queen. This shows that queens are the product of the normal genetic process, not epigenetics, which is fake science.

It is also claimed that genetic diseases are the result of gene switching by unknown chemicals from junk DNA[3.6]. In my opinion, this is quoting unlikely evidence in support of a questionable hypothesis. It is said that genes are switched on or off by *'regulating elements'* or *'regulatory proteins'*, which cause changes in gene expression. It's all pretty vague, as shown in some of the papers I have read. One paper I have read, about cellular differentiation and the role of cell signalling says, *'Interestingly, little direct data is available . . . and the majority of current knowledge consists of speculations on plausible candidate regulators',* and later: *'In summary, the role of signalling in the epigenetic control of cell fate in mammals is largely unknown.'*

Today, most of these problems with transmigrational inheritance have been hidden by highly technical jargon, while the explanation is, in my opinion, only chemical observations made during the normal process of life, which do not amount to a mechanism explaining epigenetics. As the reference says, '*the epigenetic control of cell fate in mammals is largely unknown'*.

In fact, the explanation of epigenetics is simple. Changes in the next generations are due to the normal process of inheritance, where alterations due to environment or unconscious desires are inherited until blended out. It is the lack of understanding about the nature of inheritance that has necessitated the hypothesis of epigenetics. Inheritance does not come directly from DNA, but **directly** from the parents, making the idea of epigenetics irrelevant; although some treat it as important modern science.

Stem Cells and Microtubules

Microtubules are very special cells that are at the heart and stem of all life. A paper by E M Insinna, *Synchrony and Coherent Excitation in Microtubules* (1992 France), has summarised the work and theories regarding how microtubules function and how they potentially arrange cell differentiation by cells speaking to cells (although without sufficient evidence, all theories remain just that – theories). He also suggests that brain microtubules might sometimes communicate between the members of a species. He uses the word 'synchronicity', whereas I would use words such as 'brain sharing' or 'telepathy'.

After the fertilisation of any plant or animal, cells are formed that have special reproductive properties. They repeatedly divide and are known as 'embryonic stem cells. These eventually divide to form slightly different stem cells called 'adult stem cells. The mystery here is that they either divide to copy themselves or divide into specialised cells capable of forming any part of the organism.

After fertilisation, a single human embryo cell will multiply to about a hundred stem cells in five days, gradually starting to form the basic embryo shape by dividing into ordinary cells to make up an organ, bone, or skin. This is the process known as differentiation. It is said that the mystery of this process is due to direct signals from DNA, but no mechanism has been found for this after fifty years of searching. The challenge here is that all growth and repair seems to be guided by stem cells giving information that causes cells with identical DNA to differentiate or divide into different types of cell, all in the right place, to form blood vessels, nerves, tissues and muscles. This is another example of something that is *'not fully understood'*.

Even if DNA directly guided fetal growth, this still could not explain how hair and nails grow. It is the adult stem cells that are strategically placed in every part of our body to control the correct growth of hair and nails, as well as skin and bone repair after an accident or operation. These cells are special to each organ and are few in number compared with the number of cells in the organ, whether it's liver or flesh. It seems to me that detailed information is given to the stem cell from the brain, which enables details of the organ, bone, or flesh from other members of the species to be copied.

Although microbiologists have learnt how to multiply embryonic stem cells in a Petri dish, they cannot multiply the adult stem cells in the same way, because adult stem cells are controlled from the brain. This is a problem because it is thought that great advances in medicine could be achieved by transplanting adult stem cells of the correct type, which could cure things like cancer or diabetes. The research is tightly controlled in the US and Europe, but an international team has urged China to abandon selling unproven cell therapies.

In order to examine some of the riddles of stem cells, animals have been studied that first appeared six hundred million years ago. The stem cells of sponges and some flat worms are remarkably like humans, but they are much more adaptable than ours in that they can change into different types of stem cells, enabling the regeneration of a whole organism from a small piece. For example, one species of small freshwater flatworm will regrow into about ten worms if cut into ten pieces. Chemical analysis of the stem cells has been carried out in the hope of finding their fundamental action, without success.

The goal is to identify how embryonic cells become differentiated cells that form the tissues and organs. Proteins have been identified that induce differentiation, but where these proteins come from is a mystery; clearly *'not fully understood'*.

The same applies to microtubules. These are little understood also, although they form part of every living cell, known as the cytoskeleton. They appear to be the control system of a cell that serves an important purpose in the process of growth. In the absence of a how and why regarding microtubules, I can present my own ideas, based on the evidence. So, what is the evidence?

Microtubule biology is amazing. Although microtubules were discovered in 1903, it was another century until they were properly researched. They are semiconductors, with one end negatively charged, generally pointing to the nucleus. Microtubules outside the cell (cilia) are sometimes used as waving paddles to give movement to a single celled creature (paramecium), and they also seem to give the creature a degree of intelligence. Research has been unable to show how they work. In humans, the synapses of brain neurons and nerve cells are microtubules, as are the tails of human sperm.

Our central nervous system, passing down the spine, contains millions of microtubules, each of which is a hollow bundle of thirteen separate tubes. These nerves connect with all areas of the body – two hundred of which have been identified and named. If I burn my finger, the nerves will tell my brain, which will register pain, and then other nerves will tell my hand muscles to move my fingers. Recent research indicates that the sensing elements of sight, sound, taste and smell are all microtubules.

There are two different types of microtubules. Firstly, there are tubes that bend or move. These are composed of thirteen tubules. Secondly, there are those that are static. These comprise nine rows of three, arranged as a circular pipe of twenty-seven microtubules. Basically, the thirteen-tube type has visible connections to DNA, and they move in a coordinated way to arrange the splitting of a cell during growth or repair. However, the twenty-seven-tube type (primary cilium or centriole) connects to the thirteen-tube type within the cell, but generally protrudes through the cell wall. It has been suggested that the primary cilium is in some way the brain of the cell, being critical for cellular communication in fetal development, as well as organising all cell division. Nearly all mammal cells have a primary cilium protruding from the cell membrane, which is the same for brain cells. Having said that, little else is known of the primary cilium. For instance, what it's for or how it works.

A logical suggestion for the primary cilia is that it is like an aerial, in that it transmits and receives information between cells by a quantum process. Consequently, it has been called 'the aerial'. Stuart Hameroff, who is Professor for Anaesthesiology at the University of Arizona and Director for the Centre for Consciousness Studies, suggests primary cilium allow brain neurons to speak to each other, accounting for brain operation and memory. He notes that complicated twisting in the bunches of microtubules allows spin currents to interact along the spiral pathways that transmit information from one cell to another, whether it's neurons in the brain or stem cells in the body. It seems that this quantum process is the way that brains work. This same process arranges the differentiation of all cells during the growth process.

Hameroff also suggests that microtubules are biological waveguides, and now scientists are beginning to study the details, as can be seen in the paper *Microtubules in Biological Cells as Circular Waveguides and Resonators* by Jelinek and Pokorny of the Czech Republic. Here they are only experimenting with X-rays and light; no work has apparently been done with spinning bio-particles. It is clear to me that the dimensions, twist and number of tubes in primary cilia form an essential biological resonator, or waveguide. Having worked on radar, making complicated waveguide parts, I can relate to these comments regarding biological waveguides and the fact that they are both transmitters and receivers.

The evidence is that they are not only waveguides to carry biological signals in the nerves of a body and brain, but they can also transmit signals from body to body and brain to brain, which is the mechanism of growth, giving information to embryonic stem cells, and later to fetal cells. This revolutionary extension, to the acknowledged mechanism of microtubules connecting with microtubules, works all the time I suggest across space. It connects cells to cells, body to body and mind to mind, which is of course the major factor in the growth of all life.

Let's see how this mechanism could work. Many scientists, as well as Hameroff, have decided that microtubules in the cells are, in effect, transmitters and receivers within the body. I have concluded that a signal could also pass over the biological worldwide web from a stem cell in the parent, along with the relevant password section of DNA, to be received by its similarly positioned embryonic stem cell in the growing fetus. Stem cells organise differentiation in their local area and, in this way, they can arrange correct copying by differentiating the cells under their control to correctly grow the details from the parents.

This would apply to growth and form, as well as the transmission of an operating system, instinct, and the details of inherited behaviour, etc., by brain sharing. Later, the same system would enable mind to mind after birth, where embryonic stem cells still remain in the brain.

Sir Roger Penrose, an English mathematical physicist, has investigated the science of microtubules and quantum physics in collaboration with Hameroff, and they talk of processing information between cells, but this is very much about the hidden nature of nature, otherwise known as quantum biology. These ideas fit well with the evidence of direct growth from parents, as well as brain sharing, inheritance and diversity.

New Core Theory

The science of biology went in the wrong direction seventy-five years ago, based on an incorrect fundamental theory. Many biologists accept this, but they have no alternative suggestion to bring to the table. There are several associated unaccountable mysteries: inheritance, diversity, evolution, the paranormal and, lastly, reincarnation, all with irrefutable evidence that is either ignored or rejected as fake news. My new core theory shows that all these mysteries are logically accountable by the simple expedient of looking at the biological problem in a different light.

It is obvious that we are random copies of our parents and their families, and although our DNA looks like a book of instructions, I have repeatedly suggested it only enables correct copying.

My new core theory suggests that DNA is simply an enabler for selecting the flow of information from parents or family members, which then controls the form of body and mind of a fetus. This includes things like a control system for the heartbeat and our immune system, etc., as well as instinctive programming of the brain and mind.

Some evidence for this copying process is given in the next chapter, where a flow of information is noted from one person to another. I am confident that brain to brain data transfer is essential as part of the growth process in order to install the operating system of the brain and body, as well as instinct, so it is understandable that it should continue into later life as an unconscious transfer of information, which is known as telepathy.

During early fetal growth, all members of a species are initially clones, only later inheriting family features and differences from parents, as well as incorporating their parents' own personal changes due to the environment and behaviour. These inherited differences are added to fetal growth information in the late stages of development in the womb, not as is generally thought via DNA, but governed by the flow of information received by the primary cilia microtubules. Movement of information between different members of a species is sent on a carrier, something like radio, for which there is now sufficient evidence, which I will detail in chapter 5.

This revolutionary idea answers many mysteries and at the same time agrees with all available evidence. It has been shown that primordial single cell creatures had microtubule-like structures, with the same basic architecture as modern microtubules. It is therefore reasonable to assume that life has always relied on quantum information flow by means of their microtubules transmitting and receiving information. It seems that growth information flows to the brain, which works in parallel with the local stem cells during the organisation of differentiation and growth.

Despite the best endeavours of scientists, it is not possible to conceive of any other system that controls the shape of limbs and internal organs, as well as a system for constructing the amazing mechanism of a brain, complete with an operating system and instinct. The function of DNA in every cell is to accept only the information that relates to its species and family by acting like a gateway with a lock. It seems clear to me that microtubules are able to move information from one member of a species to another, and are, therefore, essential during early growth.

A honeybee, which grows from an egg in three weeks, can see, walk, and fly, all controlled from a small brain that is already built, with operating instructions for its eyes, legs and wings. But more than that, it has instructions for almost thirty different behaviours (instincts). These instructions are freely available from the world of the honeybee by microtubules transmitting and receiving information. I am very familiar with the lifecycle of bees as I keep over one million of them.

My new core theory shows that inheritance alters the next generation, not by direct instructions from DNA, but by the direct copying from parents' form, etc, by inheritance. This enables small changes in each family's genome, which will then be subjected to Darwinian selection, thereby altering the overall shape of a species. Everyone inherits personal changes from its parents that were due to alterations in their environment, food, life experiences and desires. If those changes prove to be successful, they will be repeatedly inherited by subsequent generations. This is the reason why every family and individual is different to the rest of the species, causing what is known as diversity, which can alter the evolution of a species.

Not only does diversity alter the form of a species over time, it also changes an individual's instinctive behaviour, its ability to survive and its brain, mind and personality. It is by means of these small changes to an individual's barcode that differences become part of the future family's digital code, possibly altering the species code. This is the driving force of inheritance, diversity and evolution. As I see it, DNA stands for **D**ifferences **N**ot **A**rchitect.

By suggesting that DNA denotes only the differences between species rather than it being a book of building instructions, I am putting forth a fundamentally new way of thinking about genes, inheritance, and the barcode. I have tried to explain the difficulties and anomalies in the science of biology, pointing out that there has been no explanation for inheritance, diversity, mind, and brain, nor indeed brain to brain – also known as telepathy.

These anomalies, taken as a whole, represent a crisis that has been ignored. As the American philosopher Thomas Kuhn noted in his book *The Structure of Scientific Revolution,* where anomalies have been ignored, a crisis may lead to a new theory. I am aware that my new core theory will not replace the old theory by a proof, but the evidence and argument I have presented amounts to one.

My long and connected argument starts by showing how a single fertilised cell relies on the primary cilia to receive **only** the information that is censored by its DNA. That information instructs embryonic stem cells on how to differentiate their local cells and change into one of many different types of cell, so that growth may proceed as a copy of the parents. The same applies to brain stem cells, which grow brain as well as the nervous system to connect the stem cell network. I believe that this outline, based on the evidence, is the way life works.

My personal interest in all this started with a firm belief in telepathy and a conviction that there had to be a natural explanation for it. The next chapter is about my thoughts and experiences with telepathy and ESP.

Making Sense of Life

4. Brain Sharing: Telepathy

The whole history of science shows us that whenever the educated and scientific men of any age have denied the facts of other investigators on a priori grounds of absurdity or impossibility, the deniers have always been wrong.

Alfred Russel Wallace

I have much respect for Russel Wallace, a British naturalist who independently conceived the theory of evolution, and then wrote to Charles Darwin in 1858, who had been working on the same idea. As well as being a great naturalist, Wallace, who had seen the evidence of evolution in the jungles of Borneo and Sarawak, had also witnessed supernatural transfers of information from one brain to another. He had a firm belief in the paranormal, which earned him total contempt from the scientific establishment, which made sure it was only Darwin's name that was linked with the origin of species. Wallace saw the evidence of psychic phenomena for himself and happily went public with what he had witnessed. He demonstrated hypnosis (mesmerism) and witnessed irrefutable evidence at seances. In those days, all psychic phenomena were referred to under the umbrella term of spiritualism, although today that only refers to mediumship and spirits of the dead.

But we all have Wallace's dilemma: should we believe the stories or evidence regarding the paranormal and telepathy, and how can they be explained? I'm sure it is all part of the normal process of life, where we are copies of our parents and where information flows from brain to brain, giving us our instincts. This is the normal information flow that also gives us telepathy, but many people regard it as paranormal, or even impossible. I have realised that the term telepathy and brain-to-brain are not always relevant, so have coined the term brain-sharing, to be used in this book.

Some Basics First

The suggestion that someone could read your mind is normally regarded as ridiculous, **but** some people say there is such a thing as clairvoyance, mind to mind or telepathy. So, what's going on? Scientists say it's all wishful thinking and imagination, with no scientific evidence, **but** (another big but) it could be that they are wrong. Yes, there is wishful thinking and imagination, but for thousands of years, and across the globe, there has been evidence of mind to mind.

Today, there are tarot cards, pendulums, tea leaves and Ouija boards, all of which sometimes produce the correct results, which people find hard to accept. Also, the scientific study of mind to mind has produced evidence for thousands of successful telepathy tests, carried out mainly at universities in Europe and the US; and the existence of that evidence cannot be ignored, however unacceptable it is to the scientists.

Deliberate repeatability of psychic phenomena (psi) is almost impossible. Even then, if success is achieved, it is more difficult the next time, due to an unconscious barring of the mental procedure. Furthermore, people are generally very uncomfortable with the idea that someone could read their mind. Consequently, it is fear making us reject the idea of telepathy as absurd or impossible. Our brains automatically reject mental probing from outside, keeping us safe. But under certain psychological conditions, which can be set up scientifically, mind to mind can be encouraged between normal people.

So, could I read your mind? Generally speaking, the answer is **no**, but occasionally, under special circumstances, the answer could be **yes**. What are these special circumstances? The time of day can help, as I will show in the next chapter. Also, it can work better between some pairs of identical twins, sometimes between mother and child, and/or if the people involved share a relaxed expectation, a rapport or similar emotional thoughts.

During tests, the sender (agent) is normally thinking only of one thing, based on emotion, worry, fright or fear, whereas the receiver (percipient), who is in an adjacent room, is in a relaxed state of mind, but attentive. Obviously, states of mind are very difficult to describe and even harder to achieve deliberately. Anyone who has tried meditation will probably know what I mean.

Some psychic people have to endure occasional strange happenings during their daily life, like my friend Lucy[4.1], and they generally learn to keep quiet about it because people do not understand, not accept psi, and think you must be lying about it. Lucy is one of those who have to keep quiet and she wonders *'what the devil is it'* that could cause these strange occurrences. But some have the ability to consciously access the instinctive brain state deliberately during the day, and with training they may be able to make a living from it.

Psychologists tell us that we live in a dream state, and they are quite right. We go in and out of a light trance state without noticing anything at all. Our conscious mind is mostly memory, equipped for searching, thinking, and imagining, while our instinctive memory is comparatively small and hidden from consciousness. This is the unconscious brain. In the world of wolves, primates, horses and whales, there appears to be a process of sharing information between individuals, which is like human telepathy. Wolves have a psychic link to members of their pack; cats and dogs undoubtedly often know their owner's intentions. When we have a rapport with animals, we can sometimes communicate with them, it's called anpsi.

Two Types of Evidence

Over the years we have learned the circumstances under which telepathy or brain sharing can occur, there being two distinct sets of circumstances.

Firstly, it sometimes occurs *spontaneously* in a dream, sometimes totally out of the blue, just a feeling, a thought, a vision or even a pain; for example, a headache at the time when a close relative has received a head injury. These individual experiences of brain sharing have been experienced by people and reported since records began.

The second is when telepathy is *sought* by a medium trying to get information – called a reading – or by a deliberate experimental procedure of some sort or another, like guessing special cards or symbols (Zenner cards), or healing. It can even involve trying to guess who is calling on the phone before lifting the receiver. A good example of a live telephone telepathy test can be seen between the Nolan sisters on Rupert Sheldrake's website.[4.2]

Ninety years ago, an American lady called Mary Craig Sinclair, known as 'Craig', who felt she had a telepathic talent, started doing experiments with her husband and others. They would draw sketches in one room while she relaxed in another, trying to guess and replicate the illustrations. She obtained good results, seventy-five per cent over two hundred and ninety drawings. Craig produced sixty-five successful replications and one hundred and fifty-five partial ones. Some of the tests were done with her brother-in-law, who lived over forty miles away. No one can seriously claim that such a set of coincidences could happen by chance. There must, of course, be some other explanation.

Craig wrote about the state of mind that worked best for her during these experiments, and she later self-published a book about her experiences. The book, which is still available[4.3] attracted much interest as well as hostile scepticism. However, Albert Einstein wrote the introduction to the German edition, so some scientists were prepared to read and wonder how and why.

Telepathy, sometimes called mind to mind, is really brain-to-brain communication, which happens unconsciously, and is the same mechanism that transfers instinctive information from brains of parents to a child in the womb.

It was found that image transmission often occurred during dreams, and so for many years' tests were carried out in a dream laboratory at Maimonides Medical Centre in Brooklyn, New York. Four hundred and fifty trials showed that images could be successfully transmitted during rapid eye movement sleep (REM). The odds against chance were seventy-five million to one. This was a revelation and perhaps the start of scientific tests for *sought* telepathy. However, as it was expensive and time-consuming, a decision was made that testing during daydreaming might prove easier. By putting the receiver into a state of sensory deprivation, successful results were achieved under much simpler conditions.

This set of conditions is known as the ganzfeld, which was developed by parapsychologist Charles Honorton when he worked at the Maimonides dream laboratory.

Ganzfeld and Image Transmission

For about forty years, the ganzfeld has been used successfully to produce around three thousand positive results in the scientific study of telepathy; also called anomalous cognition. As part of the experiment, the telepathic receiver (recipient) is put into a relaxed state for about thirty minutes. This is achieved by asking them to recline under a red light with half a ping pong ball taped over each eye to give a flat field of view. I always think that just closing your eyes would be more relaxing, although you might tend to go to sleep. Headphones with a constant hiss (white noise) complete a feeling of isolation without inducing sleep. The object is to avoid the possibility of outside stimuli, so at the start of each session progressive relaxation is used in order to slow the mind and increase the recipient's responsiveness to any psi messages from the sender (agent), who is in another room.

Then, during a thirty-minute test, the agent concentrates on the target photograph, which has been randomly selected from four, and tries to think of any emotive aspects of the picture. After the test the recipient is shown the four pictures and chooses one based on his memories of the images received during the test. According to chance, he should guess the correct picture only twenty-five per cent of the time. Average receivers get it right twenty-eight per cent of the time; whereas, in tests carried out in 1979 by the late Dr Carl Sargent at Cambridge University, selected receivers achieved an eighty-three per cent success rate.

The Royal Institution in London has been presenting the latest science for over two hundred years. In 1981, I visited it to see Dr Sargent, who gave a lecture on his ganzfeld test results. Unfortunately, the lecture went down like a damp squib, although it should have blown the science of the paranormal sky high. People present, who did not believe in telepathy, viewed the whole thing with disbelief, while the believers were perhaps just bemused. Some parapsychologists rubbished his results, saying the test procedure was wrong and unsound, but I was fascinated at the time, and still am 40 years later.

Since that time, ganzfeld has been regularly shown to produce positive results using various methods and types of subject. Pairs, sharing a close friendship or emotional bond, produce the best results. A test in the US between artistically gifted musicians produced a seventy-five per cent hit rate. This was perhaps because musicians are especially creative and focused. Over the years, the procedure of ganzfeld has become more refined to avoid any further criticism of the technique. Auto ganzfeld utilises a TV monitor, with recorded commentaries to analyse the results.

But if you look up the ganzfeld experiment on Wikipedia, you'll find it suggests that it's all make believe and, to quote Russel Wallace, the 'facts are denied as absurd and impossible'. This is because

Making Sense of Life

organised sceptics have deliberately and aggressively set out to undermine the experiments, inferring that poor methodology negates all results. Sceptics have been allowed to deny the evidence by quoting only the results of failures and reports by committed sceptics. Look up 'telepathy' on Wikipedia to see what I mean.

Also see; skeptic.com/get_involved/fix_wikipedia They say; *"You can personally correct any Wikipedia article. As long as you can cite references, you can add the best available skeptical information to any article that needs it. Best of all, it's rewarding and fun to use your skeptical knowledge to enhance an essential public resource".*

Science and the Paranormal

Most scientists are unwilling to consider a concept that mentions the dreaded word telepathy. Sceptics state that extraordinary claims require extraordinary evidence, and they are right. The extraordinary evidence revealed in this book leads to extraordinary claims regarding the unsolvable problems of biology, instinct, the barcode, telepathy and the many associated phenomena of brain and mind.

The history of the paranormal is an interesting one because man has always wondered about abnormal mental experiences. The SPR was founded in 1882 to examine telepathy, deathbed apparitions and life after death. However, by about 1930, the society needed to be regarded as scientific, so it started card guessing tests and experiments, which could be quantified but produced minimal results. These tests have only resulted in parapsychology running out of steam, leading to little scientific progress or acceptance.

Thousands of records show that there can be psychic communication between people who have emotional bonds, where one has a crisis or accident. This can occur, for example, between a parent and child, a brother and sister or a husband and wife. The likelihood of these communications is subject to what appears to be a variation in the ability of the receiver of the information – perhaps it should be called psychic ability – with their level of rapport being another variable.

As an example of this sort of thing, I like to quote Nikola Tesla, the great scientist who invented electric alternating current power generation machines. He knew that his mother was ill and as a result had difficulty sleeping but upon waking from a dream, he later wrote, *'In that instant of certitude, which no words can express, came upon me that my mother had just died. And that was true'.* Tesla experienced various telepathic events during his life, generally due to the emotion of bereavement, which is without doubt the strongest emotion.

In this chapter, I have set the background to brain sharing, or telepathy (call it what you will); the reality is that these things are a fact of life. I like to use the term *Paranormal Is Normal*, which is the title of my forty half-hour podcasts, which I recorded in 2011 and have listed in the appendix at the end of this book. For these podcasts I interviewed many people, who helped to show that brain sharing is sometimes a fact of life and that 'Paranormal is Normal'.

Normally, we do not receive psychic information by brain-sharing, even though it might reside in our unconscious brain. Some people do not believe that someone could read their mind, but they would happily attend a spiritualist church to receive a message from a deceased relative. Some spiritualists will say it's all extrasensory perception (ESP) with the spirit world, whilst avoiding the possibility that it is really only sensory perception or telepathy. The only evidence for the spirit world is in people's beliefs; I believe spiritual messages are no more than ESP between the sitter and the medium.

Brain sharing covers a wide range of experiences, from the occasional intuitive flash, through to a medium's deliberate and sustained insight. Most people never have an experience, but some, like me, have occasional flashes with family or friends where there is a common mind pattern or thought. I have found I can recognise the strange feeling connected with that intuitive flash, which is a mixture of calmness, happiness, and excitement due to a rather special state of mind. To put it another way, it's a mixture of trance and arousal; and is as difficult to describe as it is to achieve.

Then there are the various ways mind sharing is unintentionally experienced, for example, during hypnosis, meditation, healing, prayer, the ganzfeld, lucid dreaming or others; but to achieve any sort of reliability is virtually impossible, except in the ganzfeld.

Success is generally linked with a personal quality, where a person's score on a scale of one to ten, varies with their mood, confidence, emotion and so on. Some are born with little ability and score naught out of ten, while others are born at seven or eight and become known as psychics. They must live with irritating insights, although they can be trained to become healers or mediums. Meditation or guided meditation can increase a score by a point or two; but the ganzfeld can add four or five points to both sender and receiver as it encourages the necessary frame of mind in both.

This book presents an integration of recent developments in the fields of biology, quantum science, parapsychology, the inner experience of consciousness and evolution which, if all are considered together, alters the science of them all. Science has advanced at such a pace that it is often regarded as unassailable, but apparent deficiencies in all the aforementioned disciplines have to be addressed to achieve a synthesis.

For me, things like healing, reincarnation and ghosts were formerly just too strange to believe, but an understanding of the science behind them, enabling a view of how and why, has changed my beliefs. For now, at least, try to put aside your own misconceptions – it's really not all make believe even though to scientists it's *not fully understood* so is best totally ignored.

I have shown that there are two types of evidence for the paranormal; spontaneous and sought. The spontaneous type is always anecdotal so cannot be regarded as scientific evidence, even though there is no shortage of reliable cases; as I have shown.

There have been many tests of paranormal phenomena, carried out to scientific standards of acceptance, but they are still unacceptable as *'they can't happen -and so they don't happen'*, which is a completely unscientific approach. The Ganzfeld is the best of example of these tests, but card guessing, sense of being stared at, dream telepathy and remote viewing are others, which have been carried out to scientific standards and have been verified by repeat testing.

However, scientists do not ask if the evidence for a new theory has been verified, but rather about its probability, in the light of the evidence that actually exists[4.4]. As the evidence for DNA being a book of instructions for life has been misinterpreted and incorrectly verified, a more plausible probability, which has known evidence as presented in this book, is an acceptable theory.

The hard facts regarding psi are firstly that it exists, but the evidence is not acceptable as there is no reason **why** it exists and there is no possible suggestion or hypothesis regarding **how** it works. And taking these two in turn; I have shown that the new theory of biology requires that inheritance flows directly from parents' families to a growing child, which is the reason **why** information has to flow across space in a similar way to our radio networks. And, as with radio, there must be a way that the information flows from person to person. There must be a carrier of that information to give a reason **how** psi is a vital part of life. A hypothesis giving the mechanism of the carrier, based on the hard evidence, is given in the next chapter.

Making Sense of Life

5. It's the Planets

All life relies on information flow between members of its species, but up until now I have not asked the question: what's the carrier?

Man has for a long time questioned how knowledge can sometimes spread from one mind into another: it is called telepathy. Modern scientists have tried in vain to measure some sort of signal, or electromagnetic carrier, when there are messages sent between two people. This chapter details an amazing voyage of discovery of evidence from many areas of scientific interest, from astrology to astronomy, particle physics, the ganzfeld and the planets. Each piece of evidence taken on its own may sound like science fiction, but when taken as a whole, they are a revelation. I had researched and considered all possible carriers for many years, all to no avail – then I had a breakthrough.

Anomalous Cognition

When attending a lecture on anomalous cognition, which is the scientifically acceptable term for telepathy, the speaker mentioned the result of an analysis of almost two thousand five hundred successful telepathy tests. He said a peak of telepathy at one particular time of day was four times the average. Apparently, that peak time varies every day by four minutes; in other words, it is not clock time but the time of the stars, known as local sidereal time (LST). Psychics had noticed some variation in the likelihood of telepathy connected with time, but it was all thoroughly confusing, changing completely throughout the course of the year.

As a result of these correlations with time, a computer analysis was made of two thousand, four hundred and eighty-three successful ganzfeld and remote viewing tests, which was reported by James Spottiswoode[5.1], a British-born physicist and the Director of Research at the Cognitive Sciences Lab in Palo Alto, California. During the ganzfeld tests, the time of day had been recorded. Spottiswoode found a sharp peak at 13.30 hours LST, with another broader and lower result, which was centred on 23.00 hours LST. The results showed conclusively that telepathy was 3.82 times more likely at 13.30 hours than five hours later. This surprising discovery came out of a twenty-million-dollar US government-sponsored programme into ESP, named Stargate (1978 to 1995).

The findings of this analysis are remarkable, especially because the time of day was only given as a minor part of the test results – at that time nobody actually knew of the LST effect. This proves that there was no likelihood that the recorded times were altered. The report gave no indication of how or why, and concluded that the carrier for telepathy must originate from the stars.

This was the breakthrough I had hoped for; scientific evidence showing that an information carrier supported telepathy, which might also explain how instinctive information is carried into the brain of an unborn child or mammal. I had been aware of the likelihood that telepathy accounted for instinct, so this was the scientific proof of a mechanism, and the origin of the carrier of information.

Could There Be Something in Astrology?

Man has studied the stars for thousands of years, and there appears to be some relationship between the constellations, planets, and human activity; something more than just horoscopes. If evidence was found in astrology that corroborated Spottiswoode's results, then it would be highly significant, and I believe this is what has been discovered.

One of the books I read about astrology[5.2], called *Cosmic Influences on Human Behaviour,* was the result of a large study carried out by the Frenchman, Michel Gauquelin. He was a traditional astrologer, who was dissatisfied with the methods and results of astrology, so along with his wife carried out his own investigations regarding planetary effects. He found that people tend to fall into one of four groups, which are linked to the four major celestial bodies. Scientists tended to be born as Saturn was rising, similarly, writers were linked to the Moon, actors with Jupiter and athletes with Mars. This, Gauquelin discovered, was determined by heredity from their parents. A similarity in personality and facial looks can also be seen in people born under a certain planet. When combined with Spottiswood's findings the result is amazing science fact, not fiction.

Gauquelin and his wife studied twenty-seven thousand birth certificates, noticing a distinct hereditary factor concerning the time of birth. Where both parents were born under a certain planet, there was a strong correlation with their children, but less so when the parents were born as different planets were rising.

They noted the birth times of people eminent in certain professions, finding that there was a strong correlation between their times of birth and subsequent vocations and personalities. He obtained the birth details of two thousand and eighty-eight sportsmen in Europe, finding the odds against chance of them being born as Mars was rising, or at its zenith, was millions to one. To Gauquelin, this was exciting, but not unexpected, as it was part of his belief in astrology. However, his book produced much outrage, which led to the results being checked in Belgium, only to be upheld as correct. This became known as the Mars effect, where the time of birth indicated a correlation between the rise of the planet Mars, with the subsequent lifestyle and profession of the individual.

This is all very strange, but it does make the crucial point that the planets emit some sort of carrier, which is unconsciously detected by the child in the womb in order to time its birth at the rise of its hereditary planet, or when it is at its zenith. It is, of course, extremely difficult to offer any reason why birth times should determine personality and lifestyle, which is the accepted idea in astrology.

Gauquelin believed the time of birth was determined not by the planets, but by the child in order to coincide this moment with the rising of its particular planet, chosen by heredity. This was achieved, he said, by the child's hormones triggering the birth process in the mother, which is the excepted process. For me, this is a reasonable explanation for what became known as the Mars effect.

Gauquelin put much work into collecting his data, which was then statistically analysed, followed by several careful replications. This makes it believable *if* the answer can be found as to *how* and *why*. The Frenchman was as mystified as everyone else concerning how a child could be aware of planetary positions.

I asked myself how this could be; how could a child detect the position of a planet? And I realised the answer was obvious, if it's accepted that a new life relies on a flow of information from its parents. For nine months, the growing child has been made aware of fluctuations in the flow of growth information. It stands to reason, therefore, that the child will seek to be born when the information flow is at its strongest, which is something of which Gauquelin was not aware, although it does answer the question of *how* and *why* a child knows when its planet is about to rise or is at its zenith.

We are all somewhat like our parents, due to our inheritance, but part of that inheritance could be the planetary type of our mother and father, regarding physical appearance, personality, and likely profession in later life. We have seen that a fetus receives data from its family, using its DNA code as a key. For this reason, a child will time its birth when the reception of data from its parents is at its best, which is the time when the carrier is at its strongest. For me, this offers a reasonable explanation for Gauquelin's dilemma, explaining *why* and *how* a child can be aware of a planetary position. It also explains how and why there is this astrological link between the time of childbirth and future profession.

Traditional astrologers have always claimed a relationship between people and the stars, but they have not been able to obtain any evidence (such as the evidence Gauquelin uncovered). It now seems that there has always been 'something in astrology'.

Some people were violently opposed to Gauquelin's results and accused him of selecting data, which in one respect was correct. He had chosen eminent members of each profession to get a noticeable result.

After years of criticism, Gauquelin tragically committed suicide in 1991. After his death, however, a re-examination of his data showed that the planetary effect was proportional to the eminence in each profession, as would be expected, so Gauquelin was exonerated of selecting data. This re-examination was carried out by a German psychologist, Suitbert Ertel, with an American astrologer, Kenneth Irving.[5.3]

While Gauquelin's reports were viewed by many with incredulity and disbelief, the sceptics were outraged. It is reported[5.4] that three sceptical organisations, the Committee for Sceptical Enquiry (CSI) in the US, the Committee for the Study of Paranormal Phenomena (CFEPP) in France and the Comite Para in Belgium, each carried out a replication of the Mars effect, in the hope the results would be refuted. In fact, each test duplicated Gauquelin's findings, which each organisation subsequently tried to hide. Eventually, several members of the CSI resigned in disgust, exposing the whole situation. Since that time, no sceptics have attempted any scientific replications.

The question to be asked now is: does this relate to Spottiswoode's report in which he shows a relationship between telepathy results and the position of the stars? Several writers have alluded to this in the past, based on the similarity of the results, but it has never been taken further. I pondered on this and realised that the next step was to investigate what planet was rising at Spottiswoode's peak of 13.30 hours. The answer is not just one planet, but billions of them.

The Milky Way

At 13.30 hours, local sidereal time, the centre of the Milky Way appears over the horizon, wherever you are in the world, or whatever the clock time is. If, as I have suggested, there is some quantum emission from planets, then there should be a large effect at this time due to the planets orbiting the two hundred and fifty billion stars in the Milky Way. It would be useful if we could make some measurement of this emission, but since we cannot we have to rely on the evidence of the effect it produces.

If we look closely at Spottiswoode's published graph of telepathy (anomalous cognition) against 24 hours of sidereal time, as well as the sharp peak at 13.30, we'll see there is another peak nine hours later that is very broad; over about five hours, in fact. This broad peak is the time the Milky Way is overhead and the carrier rains down on us, producing movement of the earth's planetary emission resulting in enhanced telepathy. This second peak in telepathy reception matches well with Gauquelin's second peak of childbirth, which is when the relevant planet is overhead (zenith).

This significant observation of the timing of two similar peaks in the flow of information carried by a planetary emission, but resulting from two completely different scientific reports, does not appear to have been noticed by anyone previously. I do believe this is the correct interpretation of the evidence. It shows that the carrier of telepathy, as well as the trigger for childbirth, is an emission from planets that is entirely unknown to science. So the question is, what could the emission be which resolves so many questions?

Spinning Particles

Having decided that the planets generate a quantum particle, wave, field or ray, I puzzled over what it was likely to be. By considering what is peculiar to planets, I realized it was the continual conversion from linear inertia to rotational inertia by gravity, I then concluded that some elemental particles of the planet are made to spin, which are released into space. This would appear to be no problem from the point of view of the power needed, because the gravitational energy needed to keep a planet like earth in orbit is about fourteen thousand million horsepower. A fraction of this appears to produce a permanent flow of spinning particles, perhaps quartz that emanate from every planet in the universe. Looking at the literature, I see a possible candidate is a hypothetical elementary particle, the graviton, which was theorised in 1930 in Russia, but has not been found or measured.

I thought that perhaps I was getting a bit out of my depth, so I researched spinning particles and the possibility that they might carry information. In an online paper by Russian scientists, Nachalov and Sokolov, detailing almost sixty years of research on spin or torsion fields, they report that spin fields can carry information, and they even include references to telepathy and biological growth. But in the West, this has all been written off as pseudoscience because some bogus reports were published to raise research money from the Russian government.

In the West, research is being carried out into the science of spintronics, which is predicted to revolutionise computers. Spintronics Info lists 13 research companies, so as I have said, the quantum carrier could well be spinning quartz emitted by planets. By way of confirmation, it has been found that DNA latches onto quartz. Quartz is, therefore, used in handling DNA in laboratories, probably because its crystals have the same double helix structure as DNA. So, could it be that DNA evolved to be read by passing quantum spinning quartz?

A paper from the Weizmann Institute[5.5] in Israel says that *'A biological molecule – DNA – can discern between quantum states known as spin',* and that *'DNA turns out to be a superb spin filter.'* It is clearly very early days in this type of research, but those comments indicate that a start is being made in studying the relevance of quantum spin in microbiological systems. I also read that you can store an almost unlimited number of bits of information in one particle.

In summary, the remarkable similarity in the timing of the effects noted by Spottiswoode and Gauquelin is very significant, even though one relates to successful telepathy and the other to birth times. This has to infer that the carrier of the information for life originates from planets, where quartz particles are caused to spin by the action of gravity on inertia.

The study of Spintronics has produced a large range of electronic components used everywhere, as well as in things like electronic gyroscopes for planes and spacecraft. But there has been no further study of DNA and spin fields, even though quantum biology has highlighted many of the problems.

The evidence is that quartz is compatible with DNA, as they both have a double helical structure. All this unlikely evidence, when taken together, is enough to show conclusively that 'It's the planets' that carry biological information for the growth of body, brain and mind.

6. The Philosophy of Consciousness

The idea that we are all direct replicas of our parents offers a surprising new understanding of consciousness, which is a result of the new core theory I have put forward.

The Conventional View

The conventional view is that the nature of consciousness is unknown and has been termed 'the hard problem', or the eternal mystery. Despite this, there is plenty known about the biology of the mind, although not much regarding how it works. It's said to be an electrochemical thinking machine, which self assembles from the data given in DNA, but there is no real evidence that this is the case.

Scientific research papers online regarding the science of consciousness[6.1] are numerous, being divided into approximately seventy subcategories and updated weekly. A few taken at random are; visual imagery and imagination, vegetative state and coma and emotion and consciousness in psychology. Despite all this, little has been found to explain how it evolved, why it works and how it grows.

Perhaps I should make it clear that when I talk about brain I mean the unconscious, with the mind being the conscious part; but the dividing line between the brain and mind is rather uncertain. Many bodily functions are controlled by the unconscious operating system that operates automatically, for example, digestion, immune system, hair growth and instinctive actions like flinching. But on the other hand, some actions are shared with the mind, for example, balance when walking, salivating, and the movement of the eyes. I believe it is the brain's operating system that decides how much information is available to the mind.

When it comes to memory, there is very little agreement or evidence to show how awareness is stored and later retrieved, but it appears likely that microtubules are in some way responsible, because each synapse, which make the interconnections in the brain, is composed of up to a hundred bundles of microtubules. Our vast memory is organised into various types that are in different areas of the brain and deal with sight, sound, digestion and instinct, etc., as well as short and long-term memory, that we call the mind. The interconnection between all these memories must be organised by a complicated operating system, of which we are completely unaware.

Scientists have suggested that the mind has at least five different types of memory:

> 1) Working or short-term memory. For example, if asked a question, the question has to be remembered whilst answering.

> 2) Semantic memory. Used when answering a historical question about who, what and why.

> 3) Episodic or time and place memory. For example, recalling where you went on holiday.

> 4) Procedural memory. For example, describing how to do something.

> 5) Perceptual memory. For example, suddenly remembering an old acquaintance.

These different types of memory are what we regard as consciousness. It's the everyday analysis of these memories that we do all the time, i.e., thinking. So that's what consciousness is: the analysis of memories.

We evolved from primates, but it was the development of speech that gave us our superior intelligence and consciousness. Scientists have examined the physical and psychological evidence in the greatest detail, while assuming that DNA provides the necessary growth and organises the data. So, the general consensus is that the solution to the nature of

99

consciousness is elusive, with no theories to explain it. I believe it's just the analysis of memories.

Most people have a basic understanding of how the mind works, but the 'hard problem' is understanding how it got there, and how it has evolved over time. A London medical doctor, James Le Fanu, writes, *'There is a powerful impression that science has been looking in the wrong place . . . there is the sense that something of immense importance is 'missing' that might transform the bare bones of genes into the wondrous diversity of the living world . . .'*[6.2]

An Extra Ingredient

Could it be that the method of growth of the body and brain, outlined in Chapter Three, in the *New Core Theory*, is the 'missing' link in our understanding of consciousness? Not only would this provide an answer as to how the brain grows, complete with an operating system for connecting the hundred million neurons and the controlling mechanism to operate the mind and body, but it would provide some guidance to the understanding of the mind. In fact, it would make the growth and evolution of mind pretty obvious.

As I see it, the creation of the brain and mind is no more miraculous than the creation of the rest of the body: everything is created by the direct process of replication. This is the extra ingredient needed to understand consciousness. Nevertheless, the control system of the brain, with its vast memory systems, is rightly regarded as the most complicated machine in the world. But perhaps now that we know **how** it develops, an understanding of it will not appear so daunting. It was the development of speech that gave us increased intelligence and consciousness. Personally, I accept my mind as an important part of me, no more of a mystery than my immune system, or legs or stomach – and the animal world is no different.

Animal Consciousness

All animals have feelings and consciousness, sufficient for their own capabilities and way of life. But what do we mean by consciousness? Scientists have not been able to make up their minds. I believe consciousness is our thoughts and beliefs, our feelings, our emotions, our fears, our dreams, and our memories, as well as our instinctive recollections and urges. Although I used the word 'our' throughout that list, it applies to the whole of the animal kingdom. They possess all of these, but, as stated earlier, only to a degree that is sufficient for their own capabilities and way of life. Horses may not have fingers or speech, but they can understand some of our words, and have enough intelligence and reasoning to thrive out in the open, despite predators, drought, wind and rain.

The brain of all mammals and birds is very similar, and they all manage to survive despite limitations in reasoning and intelligence. I recall watching a mother song thrush in my garden feeding a piece of bacon rind to her youngster. Although it was too long and stiff the mother persisted, thinking it was a worm. In the end she succeeded, but alas the young bird died because its mother lacked reasoning and intelligence.

A happier story is about the squirrels in my garden, who I feed cob nuts to during the winter. The nuts are put in a small bowl in a tree, and two years running I have been given a thank you present, set carefully in the bowl. One year it was soft pampas grass and the second time it was moss. Both were gifts for me to warm my nest through the cold weather; perhaps the most appreciated present I have ever received. I duly published a photo and article[6.3] about this offering, but some people suggested that my present had just been blown into the bowl by the wind. I argue that squirrels are more intelligent than you might think.

Another story I like is of a dolphin at sea who had a hook and fishing line fixed firmly in its mouth. It swam up to divers who removed the offending items. The grateful dolphin had realised that the men wouldn't harm it and would be able to remove the hook. These stories demonstrate that animals have consciousness, empathy, and common sense.

It is the animals that live in groups, developing group hunting skills and some verbal communication, that have evolved complex behaviour patterns leading to more intelligence; for instance, dolphins, wolves, lions and then, of course, the primates. But it was only one species of primate that learned the art of speech.

By developing communication, we have completely changed our lifestyle, and have learned to think in words, something that we all take for granted, but which puts us above all other mammals. We are part of the animal kingdom, having the same consciousness, even though it's somewhat more advanced. When it comes to other animals, they all have a brain that shows some intelligence of mind. Is that not consciousness, too?

Insects have plenty of 'how to' type instincts: for example, my bees, which number more than a million. They go out looking for flowers up to three miles away, find their own way home, and then tell their friends where they collected such good nectar by 'dancing' to give the direction and distance. When it comes to swarming time, scouts go out to find a suitable new home, and between them they might find several sites. They therefore have their own system of democratically electing the best site, by headbutting other bees while they are dancing to indicate where they have found a good location. This method of voting results in only one elected new home. Then they all fly off with the queen. It must be that way, because any bees that went off to another site would be doomed, having no queen. That inner experience is intelligence, which has evolved into a 'how to' instinct, but the method of carrying it out surely must be consciousness?

To identify predators, some species of primates have developed common calls that are like the beginnings of speech. These have altered their brains to instinctively remember and vocalise such calls. In the same way, the speech we have developed has changed our brains to accommodate this thinking and vocalising. At the same time, we have lost the skills of swinging through the branches of trees. In later stages of evolution, that same system changed our brains to allow the learning of music, writing and the internet, and so on, which explains how our brains became so different to those of monkeys. We have little understanding of sleep, dreaming or trance, which applies to all mammals.

Trance States

And here we come to a part of consciousness that I always find quite fascinating. Scientists have discovered much about trance but some of it remains 'little understood', meaning it's a complete mystery, so they ignore evidence that doesn't fit the current dogma.

This attitude is quite reprehensible. I am referring of course to the scientific view of brain-to-brain or mind-to-mind where the evidence is ignored. In this section, I only discuss light and medium trance which does not bring about mind-to-mind experiences. Deep trance, which is discussed in the next two chapters, can result in a mind-to-mind or psychic experience.

There are many types of trance, some are involuntary and some sought. They range from sleep, to laughter, crying, meditation and hypnosis. Trance seems to cover a wide range of very different occurrences which all seem to be related to a single state of mind often referred to as the dream state. Here we must be careful about the difference between brain and mind. The dream state is an unconscious brain activity that only occasionally appears to conscious mind. But we can slip into trance at any time during the day by concentrating on one thing like reading, listening to music, or even thinking of nothing at all.

A definition of trance is very difficult; some say it's when consciousness is slowed down or reduced and suggestibility is enhanced, or when there is intense focusing of attention – but that's only part of it. It's best to examine the different types of trance to see what's going on.

Firstly, there is sleep and dreaming, which have been well studied. Dream sleep is characterised by rapid eye movements (REM) that appear several times each night, during which time we analyse unresolved emotional events that have occurred during that day. This is a strange system by which the dream relives the drama of the event, comparing it with our instinctive programmes. In this way, an emotional event updates instinct in the person's memory.

In his book *The Origin of Dreams,* therapist Joe Griffin[6.4] says this is why we dream, as well as stating that we access the trance state and REM during the day when we laugh, cry or are frightened. This is the REM state being activated outside the dream state, recognized as trance. I discussed this with him during one of my *Paranormal is Normal* broadcasts[6.5], and he noted that trance is a narrowing and focusing of attention, which evolved for the expression of instincts. It seems that all instinctive actions limit the action of the conscious mind, where a part of the brain, the amygdala, senses an emotional situation causing the shutdown of the relevant part of consciousness and enabling the instinctive action to be carried out. For instance, car drivers, who become stressed or angry lose the ability to think clearly, sometimes act foolishly and with hostility due to consciousness shutting down; known as road rage. Trance and instinct are an important and fundamental part of our lives. The science is well understood, leading to a comprehension of other trance states, one of which is meditation.

Many people who practise meditation say it clears the mind and is generally beneficial. Meditation works because it empties the conscious mind and accesses the dream state, during which time mental conflicts can be analysed and resolved, making us feel more able to deal with life and any stressful situations that may come along. The various techniques of inducing meditation all aim to remove conscious thought, by repeating a word (mantra), staring at a candle, looking at a single point of nothingness or whatever works for the person concerned.

There are many methods of aiding meditation, from excessive exercise (whirling dervishes) to mindfulness (Zen), through to Western methods (T.M. ganzfeld and so on) using various techniques, such as control of breath and mind, as well as making use of instruments. All meditation is done in the hope of reaching the dream state of mind, which provides many benefits: it reduces stress, builds self-confidence, increases empathy, intuition and brain functioning, and it cures migraines and insomnia. It is also free **if** you can meditate deeply enough. I went to classes for twenty weeks and realised I was a useless meditator, only reaching the correct state on one occasion. Meditation is very much a matter of degree; most achieve only light or medium trance.

The state we generally associate with trance is hypnosis. This is no different to meditation in that we can meditate into what is known as self-hypnosis, by which we can deliberately insert commands into our own unconscious brain – but the technique is very difficult to learn. A far more effective procedure is for someone else, a hypnotist, to talk us into accessing the dream state to implant suggestions into the hidden part of our mind, or even to retrieve forgotten memories. One method is to suggest that a noted action will be carried out in response to particular stimulus (post-hypnotic suggestion). For example, it is sometimes suggested that a cigarette will induce nausea in which case the thought of a cigarette triggers the instinctive response mechanism to remember feelings of nausea.

Hypnosis generally requires only a shallow level of trance. If it is too deep there is a chance of problems, as German doctor Franz Mesmer noticed in the early days of hypnosis. Some of his subjects fainted and some wept or became delirious. In the same way, stage hypnosis can be dangerous. I once had a student lodger who made a fool of himself during a hypnotic show, when he was told he was a typewriter. That night he was found sleepwalking round the house and had to be led back to bed. Sleepwalking is the result of hyper arousal during deep sleep, and it can have terrible consequences like falling down the stairs.

Hypnosis is sometimes regarded with fear. We don't like the idea of another person taking charge of our mind, and there have been instances of unscrupulous hypnotists taking sexual advantage of their subjects, or stealing their money. Unfortunately, this fear of hypnosis tends to inhibit its success because the subject is reluctant to sufficiently relax, or sleep, as urged by the hypnotist.

Many trained therapists use hypnosis for treating various psychological problems, although they call this guided imagery. Joe Griffin and his colleagues have trained thousands of therapists to use this technique, which treats anything from sexual problems to obsessive-compulsive behaviour and post-traumatic stress disorders.

Is there anything we can learn by examining the psychological aspect of the dream state while awake? It occurs due to a relaxed mind and brain, without being asleep, even though hypnotists urge their subjects to 'sleep – sleep'. The key psychological feature aims to produce a no-self condition by suppressing feelings of self. This is normally achieved by deliberately relying on someone else, either real or imagined. The hypnotist's subject mentally submits to the hypnotist, clearing away his personal self and consciousness, which generally brings on a hypnotic state. As I mentioned previously, the self-meditator often has great difficulty in achieving a sufficiently deep state because he consciously tries to lose his personal feelings of 'me'. Healers generally say they are not responsible for the healing they are merely a channel, which is the key to suppressing any thoughts of self.

In summary, we see that during trance we are accessing the dream state, made obvious by the appearance of REM, and brought about by a variety of situations, either deliberate or accidental. During the day, unconscious instinctive thoughts are sometimes brought into consciousness by trance, but during deep sleep, an emotional experience of the day is compared with instinctive memories, guiding the imagination of the dream. This process alters our own instinctive memory, which is then inherited by our children.

The scientific understanding of instinct, trance and the dream state has added greatly to the overall understanding of consciousness, but it has nevertheless been treated by scientists as part of the unfathomable mystery of consciousness. This is because it has never been appreciated that there is only one human brain, repeatedly replicated, but modified slightly by differences in tribe and family.

My theory of inheritance shows that by adding a new emotional experience to our instinctive memory it updates our personal memory which is then subsequently inherited by the next generation. This new observation, I believe, is the mechanism by which all animal instincts have formed during evolution.

It is only by looking at the relevant evidence in a completely new way, beyond accepted theories, that the mystery of consciousness has become clearer.

Making Sense of Life

7. The 'Spooky' List

This chapter looks at some of the observed facts about the psychic side of life, and how those facts are all initiated by trance. I'll examine what we know about the way the mind changes during meditation, hypnosis and other difficult-to-understand parts of consciousness as well as revealing a scientifically acceptable explanation for all psychic happenings. Consciousness is excluded by the trance state allowing unconscious mind sharing which is essential to my new core theory of DNA. So, by examining the workings of trance, I offer the explanation for mind-sharing and psychic states.

Trance Studies

Consciousness and trance are identified with rapid eye movement (REM) when in the dream state, with emotion and instinct. The way these three interconnect is now largely understood by the science of psychology.[7.1] The main result of these interrelationships is that consciousness is switched off while we dream at night, but during the day free will dominates until something makes us laugh, something emotional occurs, or even a tiger chases us, in which case the trance and dream state reduces mental analysing in order for instinctive thoughts and actions to appear.

Science has uncovered many details of what's going on in brain and mind and we do now have many clues to help with the psychology. This section now gets a bit technical, but it does reveal how and why deep trance with REM gives explanation for psychic phenomena (psi).

A good indication of what's going on in the brain can be shown using Scientific Instruments. We can measure voltages by placing pads (electrodes) on the head; and we can measure mental arousal from electrodes placed on the palm of the hand that measure changes in electrical skin resistance of the hand (ESR) due to sweating on the palm. Although the scientific findings relating to various states of mind were published long ago, there has been little recognition, only rather grudging acceptance. Looking back at that work, we can see that it was an important scientific study, recording evidence of what happens to brain rhythms during hypnosis, spiritual healing, ecstatic enlightenment, mediumship, and other paranormal events, which all take place during trance and REM.

For many years we have known that the frequency of brainwaves corresponds to the state of mind during the trance, measured on each side of the head using electrodes. Normal everyday thinking is the beta frequency (13-30 cycles per second, or Hz), but if we relax, close our eyes and think of nothing in particular, some alpha frequency will normally appear (8-13 Hz). A slower rhythm is theta (4-7 Hz), which is the dreaming brain frequency, or when we access the dream state when we're awake. Delta (1/2-4) is the deep sleep condition.

These were the frequencies noted thirty-seven years ago by Max Cade[7.2], whose meditation and healing classes I attended for twenty weeks, but today there are wider ranges quoted, although these four basic bands remain the same. Max Cade noted[7.3] that deep sleep, dreaming sleep, half asleep and the waking state are only accompanied by a single peak in the four frequency bands – delta, theta, alpha and beta, respectively – but he also discovered that when a person accesses the dream state with REM, he produces symmetrical voltage readings on each side of the head (bilateral symmetry). Sometimes meditators would simultaneously produce three distinct symmetrical bands of beta, alpha and theta when meditating, called the triple peaks, as shown on a special electroencephalograph (EEG) machine which Cade designed especially for this purpose (called the Mind Mirror) in which rows of LED lamps showed the voltage on each side of the head over the whole range of frequencies.

The questions that have not been answered are why we have balanced symmetry, why the frequency of alpha reduces over about ten minutes and why we see triple peaks. My own answer to these questions[7.4] is that a relaxed and empty mind has a scanning system searching for thoughts to process, rather like a bird flying back and forth looking for food, and this produces a measurable voltage at alpha frequency on each side of the head. Then, after a few minutes, the cells tire and slow down, producing a lower frequency that is the same on each side of the head – the balanced symmetry. The reason we see the three peaks is because different parts of the brain oscillate at different frequencies when in trance.

Max Cade observed the trance brain-waves of hundreds of people making careful measurement of the frequency of alpha when the subject displayed balanced symmetry. Normally, when meditating it is 10 Hz, but decreases with five years of meditation practice to about 9 Hz, and 7 Hz after 20 years. He also noted that the normal alpha rhythm for children is 4 to 7 Hz, so we can see the actual frequency of alpha is pretty variable.

The triple peaks (beta, alpha and theta) indicate what's happening in the trance state, and they are in complete contrast to normal life which is the everyday beta state that shows a wildly fluctuating pattern changing as we think about different things. Cade's discovery was that only when accessing the dream state of trance and REM does the triple peak occur, and balanced on each side of the head.

Brain arousal is the other factor that alters the state of mind and is easily observed on an ESR, which Max Cade's students used during his meditation and healing classes. By watching the needle movement on the ESR, they learned to deeply relax, a process called biofeedback, which was very effective. Nowadays, the instruments used for measuring emotional arousal are known as EDA (electrodermal activity), but they are the same.

There is one aspect of Cade's classes that the students were not aware of, and that was the unnoticeable flickering lights, which he gradually reduced from an alpha to a theta rhythm (10-6 Hz). The flicker aided the symmetry of brain voltages, as well as increasing the amplitude of alpha as it reduced in frequency, all of which deepened the meditation by inducing the natural mechanism of sleep.[7.5] Strobe lighting affects the brain, especially the brains of epileptics. Max claimed that many epileptics who attended his classes received considerable benefit. Today, the mind mirror technology can be used with a home computer programme. Also, a different method of inducing low alpha is sometimes used today, making use of sound (a binaural beat). What's clear is that we can control our states of mind, but with difficulty.

As previously mentioned, the practices of meditation, hypnosis, healing and mediumship are all intertwined, though each is identified as being quite separate. In fact, they all seek to access the dream state, but in different ways. However, without the triple peaks nothing significant will happen during any of these procedures. It is known that meditation can be induced in two different ways, mindfulness or the relaxation response, but in practice both have endless variations, and the triple peaks is seldom achieved. Hypnosis requires rapport, while avoiding deep trance and the triple peaks which is unnecessary for hypnosis. On the other hand, healing requires deep trance of both parties with the triple peaks to induce psychic ability to both identify the problem and instigate a cure.

It seems that here we have been only looking at a single state of mind, by which all these rather spooky things are achieved. Not only can prayers be answered, but meditators also sometimes achieve great benefit, as do the hypnotist's subjects and the healer's patients. And if that wasn't mysterious or 'spooky' enough, brain sharing and mind to mind can happen at any time during this state of mind. You will note my use of the word *this* state of mind rather than *these* states of mind. In this state the triple peaks brain sharing is possible which is responsible for the many psychic states discussed in this chapter and the next (The more you look the more you see) which is all due to the new core theory of DNA, that depends on brain-sharing.

Musical Expression

Another topic that can benefit by examination from the aspect of trance is that of the subtleties of musical performance and expression. This is something we all feel and appreciate, but something of which we are only unconsciously aware. It's hypnotic.

When at school, I became interested in the jazz of the 1920s after listening to live broadcasts by the Glen Miller band. Something special about the music fascinated me and I was soon hooked, just as young people are today, spending too much time listening to popular music, all of which stems from those same roots. Later, I taught myself to play an instrument and with difficulty read the second trumpet parts in a local dance band, which led on to playing jazz and eventually leading my own band. But there was always something wrong - sometimes the music would sound right - but generally not. I found that a dissociated state of mind helped, but after more years than I care to admit to, I discovered it was unconscious musical expression, rubato, or what I like to call syncopation.

It was while listening to Ella Fitzgerald that I first realised what was going on. Sometimes, she was ahead, and sometimes behind the correct timing, but almost never exactly as written. She divided the music up into phrases and used a standardised system for altering the timing of each phrase. This system is used throughout jazz, gospel singing, rock, and all the different types of popular music. But what is interesting is that there are two very distinct types of syncopation, both generally used in a single piece of music.

So, what is the difference between the two types of syncopation, and how do they differ from strict tempo music as played by reading sheet music. The tune must be broken up into phrases, which can be anything from two notes to perhaps ten, and then altering the timing of each phrase by one of the two types of syncopation. I believe the Beatles referred to 'up' and 'down' as the two types, but I prefer to use 'early' and 'late', as being more descriptive. With the 'early' type of syncopation each note is early except the last in the phrase, which is late; and with the 'late' type, each note is late except the last, which is early.

To understanding their formula, try listening to an early record of the Beatles. The songs were written and rehearsed very carefully, complete with syncopation, which changed every phrase or two. It is because their syncopation was so accurate, well-rehearsed and fitted their songs so well, that their records swept the world. The fact is that all popular singers and musicians today, rely almost entirely for their success, on the use of this characteristic, habit, mannerism, or deceitful device - called it what you like.

Syncopation can only work where the singer or instrumentalist has a very good natural sense of timing, so this applies to all the musicians, in the band, group or orchestra. Some can create syncopation perfectly well, unconsciously, but if tense or embarrassed, at a special performance or broadcast, they lose that vital trance state. One early recording of Frank Sinatra has none of the effect he mastered so well later.

If the leader of a jazz band uses good syncopation, then others will often unconsciously follow suit. The first time I played in public with my band, consciously using syncopation, it gave me a feeling of pleasure and confidence. Afterwards, to my great personal satisfaction, someone came up to me and said, 'Richard, you've got it'. He clearly had no idea what 'it' was, but he made me realise that this system was not just all my imagination. Forget the ideas of musical notation: four equal beats in the bar and all that. I had demonstrated to myself that there is a system to jazz music, which can be consciously put into practice – although it's very difficult to achieve without the right state of mind.

Unfortunately, many music critics, disc jockeys and bandleaders, know little of it. We hear musicians who realise that they can mess about with timing, to enhance musical appreciation, but they get it horribly wrong. In England most amateur jazz musicians, opera singers and light entertainment singers get away with incorrect syncopation. The result is inferior and many of their listeners feel there is something wrong, not knowing where the problem lies. The performers realise this themselves. Some learn that alcohol or drugs help by removing the inhibitions that discourage unconscious automatic syncopation. This is one reason why many good and famous musicians or singers, sadly end up with an alcohol or drug problem. But I have worked with good jazz musicians who quite deliberately put themselves into a special dissociated mental state, or trance, before starting to play. By consciously regressing to a childlike trance state all inhibition was lost. Under these circumstances the unconscious brain took control of the subtleties of timing to syncopate the music.

So, what can we learn from this? From the foregoing it can be seen that syncopation is without doubt a form of hypnosis, holding the interest of brain, not the mind.

When listening or playing music there is normally only a low level of trance, with psi being unlikely. But playing with syncopation requires focusing the mind on the music which raises the level of trance; so, when warming up to sing, or play, the trance has to increase. This increases the possibility of a psychic experience. Some years ago, I personally experienced this phenomenon.

I have always enjoyed playing harmony and can do it without thought. One of the few psychic events in my life happened when my band was tuning up. A member of the band suddenly started to play a tune, and without thinking I joined in that same tune, in time, in the same key, in harmony and from the first note. After a couple of bars, he stopped and we looked at each other, somewhat amazed. It is well known that the rapport between musicians is sometimes much more than just the enjoyment of the music.

Psychic States (psi) and Brain Sharing

It's the state of deep trance with the triple peaks of brain activity that opens the possibility of brain sharing. It starts in the womb when a fetus, who spends much time in REM, receives the necessary information to grow body and mind, together with its operating system and instinctive memories from its species and family. This new way of looking at the evidence gives explanation to telepathy and the psychic states during life, even though scientists and others say, 'it can't happen so it doesn't happen', only because they have not learned how it could happen.

Healing, often referred to as spiritual healing, is the first on my list of trance states that rely on mind-to-mind transfer of information. It clearly is essential that the healer can sense the problem with his subject, followed by transmitting the healing information to be received by his subject. This clearly requires them both to transmit and receive brain to brain information.

The key to healing is deep meditation of both the healer and patient, incorporating a strong sense of oneness. Contact with the healer's hands can help, and there has to be faith that the procedure can work. This faith can be religious but simply a belief in the procedure or healer is quite sufficient. I believe that even telephone healing can work, where the healer is strongly psychic with a cooperative patient. But this is difficult to achieve and, as far as I'm aware, there has never been any scientific verification. Meditation with self-healing can be effective with psychosomatic illness, and maybe even with more serious conditions. The recommended procedure is to quieten the body and mind and focus one's attention on breathing for a few minutes, followed by positive-emotion exercises using imagery related to wellness. Call it self-hypnosis if you like, but it is self-healing.

Deep trance is the REM state which is linked to psi states and dreaming. There has been much scientific study into REM during night-time sleep. I see that 'the precise function of REM sleep is not well understood[7.6] meaning it's another mystery. But Joe Griffin has shown that during dreaming, our daytime experiences are programmed into instinct. That's part of it, but it has also been found that the REM state is vital to life. REM deprivation in humans is very detrimental to health, and will kill a rat in five weeks.

Furthermore, a new-born child spends more than 80% of sleep time in REM; so, it is clear that the information for the development of body and brain is received during REM. Also, I believe that information is transmitted by others who are in REM, making the point that the REM state is responsible for transfer of brain information to a fetus or child to aid its growth. This of course is the mechanism of mind to mind from parents to child which continues occasionally throughout life, which is clearly an important revelation and is part of my new core theory. By seeing REM in a new light, it can be seen why it is only during REM, trance and the dream state that psi occurs, giving an explanation for the paranormal which has always been a mystery.

Max Cade was asked to assist with some experiments with several well-known mediums, in order to examine their brainwaves. He found that, at first, the mediums showed the early stages of alpha, which increased in amplitude and was accompanied by theta, the triple peaks, but with the normal state of low arousal. Then, as soon as the control or guide began to communicate, their low arousal changed to hyper-arousal. The retention of a deep trance state is necessary for a medium to achieve ESP, together with high arousal to enable the discussion and analysis of the ESP messages. I ask: Are these messages ESP with the spirits, or are they merely mind sharing?

Cade carried out sixty series of Zenner card guessing tests to see if the biofeedback methods used in his classes would improve psi ability. Tests were carried out after six different levels of meditation. The results were very interesting showing the variation from chance level of 25 as follows;

1) Normal waking state (just arrived in class)   24

2) After 10 minutes of listening to music   25.7

3) After either 5 minutes of autogenic relaxation
exercises or the empathy exercise   27

4) After 10 minutes of led meditation   28.5

5) After hypnosis or group-mutual hypnosis   29

6) After 5 mins of relaxation sequences, 10 mins
of led meditation and the empathy exercise   32

The fact is that most series of the Zenner card guessing tests result in failure. But Cade was able to produce far above average results, deliberately accessing the psi state, by employing the biofeedback and methods normally used in his classes. For me this is a scientifically acceptable explanation for all psychic happenings, but I'm aware that it's not acceptable to many.

Meditation is normally not deep enough to allow mind to mind but it is the mechanism used to initiate other psychic events. Historically, there have always been reports of psychic events associated with deep meditation and healing. Meditation training aims to reduce EDA skin arousal but also achieve the triple symmetrical bands of beta, alpha and theta, which are observed when using a Mind Mirror. Cade noted that many meditators who thought they were properly meditating were '*only bordering on its edges*'. It is sad that most meditators are unwittingly 'only bordering on its edges', but the real benefit of meditating is there, only if the right state of mind is reached.

Prayer is meditation, in which emotional questions are asked for family or acquaintances, who sometimes receive and unconsciously act on the request. In this way prayers may be answered.

A different form of meditation is known as automatic writing in which case the trance mechanism allows someone with pen and paper to unconsciously write thoughts from the unconscious brain. This can be in the form of answers to questions or the retrieval of lost memories. It's a rather spooky form of self-hypnosis.

The Ganzfeld is a mechanism used to demonstrate telepathy where the sender, in trance, concentrates on the emotional aspects of a picture chosen randomly and from four possible pictures. The receiver is subsequently shown the four test pictures and asked to choose which one appears familiar to him. This procedure does result in better than chance results proving that brain-to-brain, mind-sharing with telepathy can be shown to exist using a scientifically acceptable procedure.

Over the years there have been many reports of dreams revealing information about something that has happened or is going to happen. Presumably, the information comes in a dream, brain to brain from another person who is dreaming about their thoughts or intentions. I generally regarded suggestions of precognition as being information passed in this way; rather than it being due to seeing into the future.

Lucid Dreaming is the name given to dreaming in half-asleep half-awake state where people can control their dreams. I spoke with Beverley[7.7]who attended Stanford Dream Laboratory because she was a good lucid dreamer. But the point I want to make here is that her dreams were often telepathic. She even entered a telepathic dream competition. Sometimes, her dreams revealed someone else's thoughts or intentions; Beverley thought these were evidence for precognition as visions she received were sometimes played out a few days later. This shows not only the cause of precognition but shows a connection between psi and dreaming. Her psi experience led her to be successful at healing and to realise that the human race has only one mind, or brain.

The use of mind-altering drugs will also lead to psychic events, as well as being a mental anaesthetic or stimulant. There is little evidence of direct brain-to-brain or telepathy although there are many reports of strange, sometimes frightening, events. A friend of mine, who uses marijuana regularly, said he had 'seen the light'; generally referred to as mystical experience or an enlightenment.

Mysticism - The Awakened Mind

Consciousness has many mysteries, the strangest of which is the mystical experience. It's known by about 24 different names, so why are there so many names for what appears to be the same thing, most belonging to different religious faiths? Max Cade called it the Awakened Mind as it was a good description, Richard Bucke called it Cosmic Consciousness in 1901 which was his name for the experience. My own encounter was during one of Max Cade's classes when a guided meditation suddenly led to a dream I shall never forget.[7.8]

A mystical experience or enlightenment can occur with no apparent cause. Various studies have shown that people who are stressed, worried or anxious are more likely to experience enlightenment than others, although sometimes the trigger can be a rhythmic stimulation. Whatever the trigger, the experience is always said to be impossible to mistake, as it's so overpowering and illuminating – clearly some very special state of mind. There is a feeling of total transformation that brings a joyful sense of knowing that all is good.

In her book *The Relevance of Bliss - A Contemporary Explanation of Mystical Experience,* the writer Nona Coxhead states[7.9] *'Of course, there is no way to prove this experience exists, except by having it, but the body of evidence over the course of many centuries of spiritual accounts is surely as well grounded as any phenomena studied by science. It has always struck me as extraordinary that it should be overlooked, almost singularly left out of all imaginative modern research. It is as if by ignoring it, it would simply evaporate . . .'*

She was absolutely right. It is ignored since there is nothing that can be said, because 'how' and 'why' are both mysteries, although it is apparently triggered by an increase in emotional arousal from the dream state.

At one of Max's classes, I met Nona, who had co-authored *The Awakened Mind*. In her own words, she had been 'mystically inclined since childhood', and she found Max's class overpowering. She started sobbing and had to be quietly led out.

My bliss was induced by learning to meditate by biofeedback, coupled with strobe lighting and guided meditation, but there are other ways. Speaking in tongues or praying in tongues (glossolalia) in church services will sometimes trigger enlightenment[7.10].

I attended a church in London for a while, where glossolalia was sometimes practised. The priest would invite parishioners to come forward to be blessed, then he would speak in tongues, whereupon the entranced person would keel over and fall to the ground as if having a mystical experience. I had this done to me, in my own home, by that same priest during a bible study meeting. He started speaking in tongues while I was standing to be blessed, with his assistant standing beside me. I felt the assistant put his foot behind my heels, so I naturally lifted one foot over his to support myself. The priest, who shall remain nameless, pushed my shoulders in order for me to fall backwards, but I was suitably braced by a well-placed foot. The priest and helper exchanged glances, before the blessing was hurriedly concluded. Charismatic priests have always got away with this hoax entrancement because the public will accept that the priest is able to put a subject into deep trance. Unfortunately, this sort of thing just adds to the mystery and misunderstanding of trance, enlightenment, and mystical experience.

On the other hand, most charismatic priests will give a blessing to a person who will lose their balance in trance, to be helped in their fall by nearby members of the congregation. The priest may, or may not make physical contact, and there is certainly no foot behind heels, as there was in my case.

It is well-documented that the panic state of hyper arousal can bring on enlightenment. I once spoke to a helicopter pilot, who was flying alone when the rotor hit a pylon, causing him to crash into a lake. While descending, he could do nothing, and his state of panic induced a classic enlightenment. He did his best to understand this mysterious experience, but no one could explain what had happened. Similar experiences have struck people while being attacked by wild animals. In addition, many war veterans claim to have experienced an enlightenment, as have people taking DMT. Unfortunately, many of these have been termed a near death experience (NDE), which they are not, even though the reports of an NDE are strikingly similar and are the result of the same mental occurrence.

Blissful mystical enlightenments have always been part of the human experience, and they are often encountered during religious worship. However, unknown triggers, or drugs, can also cause them. The Religious Experience Research Centre concludes that thirty per cent of humanity has the experience some time during their life, others say it's more like five per cent, so it's reasonable to assume a figure somewhere in between. Whatever it is, one of the triggers for a person in trance can be a flickering light. It is said, a man resting under a tree had a blissful enlightenment because of flickering sunlight shining through the leaves blown by a breeze. He subsequently started a religious movement.

Mysticism has very many names, so the waters are extremely mudded. Illumination, Ecstasy, Bliss, The Inner Light, Near Death Experience and Satori are just a few. When anyone asks **how** or **why**, it goes very quiet. I looked at the evidence years ago and came up with a possible answer. So, what is that evidence? Here we have countless historical reports of a sudden blissful illumination where memory is increased, life's problems disappear, and death is not feared, but the experience is seldom repeated or forgotten. My studies of trance, and the evidence of my own ecstasy, have helped to come to a likely conclusion about **how**. The essence of my findings start with the facts concerning the deep trance state, where the brain suffers an overload; apparently a damaging overload which affects the brain, cleaning or purging old cells, or opening up new pathways in the brain; a purgation. I experienced a strange headache for a few days, as have others, confirming the idea of a purgation, explaining how the brain is affected while supporting the changes reported after the event. A repeat purgation would obviously be unnecessary.

To find a reason **why** this happens we must look to human evolution. Around 70,000 years ago our brains grew, due to increased consciousness, in a skull of small size leading to problems which was solved by increasing the efficiency of our brains. This was done by evolving a mechanism to clear out old cells or making new pathways; a purgation, resulting from a brain overload. I believe this logical account of how and why we have an awakened mind could prove to be correct.

Paranormal is Normal

A few years ago, I was persuaded to make a series of internet broadcasts. I say 'make' because they were very much homemade, not scripted or even planned, just a telephone or Skype conversation with an expert in their field, resulting from a prior invitation. My editing expertise was minimal, as were my interviewing skills, but I was always amazed at how my guests proved that strange and sometimes psychic things were quite normal to them. I was able to chat to some amazing people. You may care to listen, and be as fascinated as I was, but I'll give details below of some of the most interesting ones from my appendix.

I interviewed the neurophysiologist Peter Fenwick[7,11] about NDEs, which he has studied for many years in the course of his job. He expressed surprise that the reports of people who have been resuscitated from clinical death are very similar to mystical enlightenment experiences from triggers unconnected with stopping breathing. For me, it is apparent that they are the same. We know the panic state is one way of triggering an enlightenment, so what could bring on a mental panic state better than the brain dying due to no blood supply, or perhaps the panic state of resuscitation?

But what really fascinates me is that patients are sometimes aware of things that have happened during the perceived period of clinical death. My opinion is that this is due to mind sharing with a member of the team attending them, where details of conversations are shared, or someone present witnesses a view of the patient from above. That view is what is known as an out of body experience (OBE). A Dutch surgeon, Pim van Lommel, with whom I have discussed this, has made studies in Holland, having lectured and written extensively about NDEs and OBEs. He says, '*You don't need a non-functioning brain to report an NDE*', and '*An NDE, or mystical experience, can be triggered from meditation.'* So, it seems that any mystical experience is sometimes called an NDE. This has muddied the waters considerably; for example, in the US, over four thousand cases of NDEs have been reported, but very few of these were anything to do with a death experience. I prefer that the phrase NDE is only concerned with resuscitation from clinical death.

Another recorded conversation was with Dr Jude Currivan[7.12], who was once the group financial director of two major international businesses, and has a master's degree in physics, specialising in quantum physics, as well as being a gifted healer. She believes in the connectivity of consciousness with other people during dreams or clairvoyant visions, which she has had all her life. Having a supercharged intuition in business has apparently paid off, leading to a career as a successful financial director.

I spoke to an ordained minister about praying in tongues.[7.13] He talked about the beauty of tongues and how it helps relieve the emotions and tensions of bereavement. I asked about ecstatic enlightenment due to speaking in tongues, and he quoted a lady who suddenly experienced glossolalia, which led to a mystical experience while doing the dishes at home. He went on to talk about many of the religious healing experiences that had taken place in his church over the years.

One discussion I broadcast was with Bob Charman[7.14], the Chairman of the Confederation of Healing Organisations in the UK. He spoke of how Max Cade organised a demonstration of healing, where both the healer and subject were hooked up to Mind Mirrors in front of four hundred people. The healer's triple peak brainwave pattern was transferred to the patient, in what he called 'Psi mediated brain synchrony'. For me, this is another example of brain sharing. Also, the University College London used healers for cancer work. At a hospital in Birmingham, they were used to treat irritable bowel syndrome. Charman says that there are one thousand two hundred healers working in the UK.

On another occasion, I rang Beverley, who worked for Stanford University's sleep laboratory, in connection with her lucid dreaming[7.15], which is when you are in REM but are aware that you are dreaming. She said it started when she was very young, because of falling down the stairs, and that it continued into later life, which is somewhat unusual. (It has been found that many people can be trained to be lucid dreamers by means of a device that flashes lights during REM sleep.) Beverley also discovered that her dreams could be psychic: in other words, she could mind share with other people. She further discovered how to heal herself during dreaming, leading on to the ability to heal others from a lucid dream, which is the usual method of healing. Her different experiences clearly demonstrate that psychic ability is a very broad talent. It also shows that psychics, as children, very often experience a physical, electrical, or emotional shock, which triggers their abilities.

Phil Young[7.16], an Irish hypnotist/healer/author of Mastermind International, who has studied various types of healing from around the world, has concluded that they are essentially the same. It didn't matter how far apart the healer and the patient were; the only thing that mattered was that there was a mental link between the two. He calls his own method Trance Body Healing, where the common trance state connects and heals the body. When I asked him about trance, he had difficulty putting it into words, but he called it *'the play of consciousness', 'the shifting state of awareness'* or *'a non-verbal exchange of information'*. All these I understand as brain sharing; the mechanism of instinct. When it comes to names for the information carrier, he mentioned Prana, Qi, Mana, life force, animal magnetism and spiritual healing, and considered them all to be fundamentally the same.

In all the above states of mind, I have discussed the two important common elements, which are trance and the sharing of information from person to person or via brain sharing. It seems that some of the people I interviewed also had other abilities, including healing, demonstrating the fact that there is a common mechanism involved in all psychic phenomena. From a scientific point of view, I see this as a common mechanism, but the people concerned may have alternative explanations for their multiple abilities.

This chapter has shown that there is commonality between hypnosis, healing, meditation, dreaming, mystical experiences and mediumship, all of which require the trance state and can produce that mysterious mind to mind, or brain, sharing. Deep trance is responsible for many other – even stranger – functions of the human mind.

Making Sense of Life

8. The More You Look – the More You See

Psychical research has clearly not drawn a blank. It has, on the contrary, discovered something so big that people shear away from it in a reaction of fear.

G N M Tyrrell

I'll now try to put a different slant on how these various aspects of consciousness can be viewed, which to most people appear as 'freakin bizarre, man'. They certainly are bizarre, but that doesn't mean they cannot be understood. If we understand the science, relating to trance and brain sharing, then they become believable.

Ghosts

The best book on ghosts was published by G N M Tyrrell in 1943.[8.1] He presented sixty-one cases involving apparitions and haunting ghosts, which he took from the collection of the Society for Psychical Research (SPR), and which spanned the previous sixty years. He analysed and commented on these cases and, years on, his book is still recognised as being the best in its field. There are many aspects to ghosts and

141

hauntings but, unfortunately, they are collectively regarded as unbelievable – a bit of a joke.

The scientific study of ghosts began with the census of 1889, in which seventeen thousand people in the UK were asked, *'Have you ever had a vivid impression of seeing a living being, which was not due to any external physical cause?'* One thousand six hundred and eighty-four people (9.9%) replied yes. The same question was put to a total of twenty-seven thousand, three hundred and twenty-nine people in France, Germany and the US, to which a further three thousand, two hundred and eighty-six replied yes (11.96%).[8.2] Sceptics have always tried to deny any evidence for ghosts and, perhaps without any explanation of how or why, this is not too unreasonable. Before I understood brain sharing, I used to think the same myself.

Tyrrell's analysis of sixty-one cases led him to divide them into four major classes[8.3] noting that the classes are not well defined, also some casas do not belong to any class.

He mentions experimental cases in which the apparition was sent to a particular person, and noted 16 of these cases that had been reported. An example of this mechanism, which is used today, is where in Ganzfeld a scene on a photograph is deliberately sent to a person in trance; and it works.

His second class of cases is when someone sees a recognised person who is in a crisis situation. Lucy, a friend of mine, saw her husband in a car crash and phoned his office to see if he was there. Later, the husband called to say he had been involved in an accident. This sort of case is known as a crisis case.

Another class is that of post-mortem cases. After the death of a recognised person a haunting apparition is sometimes seen which is the classic ghostly vision. Nikola Tesla, the inventor of electrical machines, knew that his mother had died from information received when he was asleep, but he might have seen a vision of his mother – the mechanism is the same.

The fourth class is where there are ghosts that habitually haunt certain places like old houses or airfields where there had been aircraft crashes. Modern ghost hunters say that ghosts are due to emotional memories becoming trapped into the stone of buildings, which are subsequently picked up by a person in trance. This mechanism is known as a stone tape recording.

However, Tyrell notes that the classes are not very sharply defined, as many cases span two of them, indicating a common mechanism.

One aspect of ghosts or apparitions that caused much discussion was when two or more people saw the same figure (collective percipients). I see this as two people in trance sharing, by telepathy, the same details of the ghostly emotion. Tyrrell suggested that collective cases were due to the ghostly visions spreading by 'infection' between the percipients.

Since Tyrrell wrote his book, ghosts have not gone away; there is a steady flow of reports, although most are not brought to public attention. For example, my friend Lucy, who had a crisis apparition of her husband in a car crash, was only reported on my *Paranormal is Normal* broadcast.[8.4] This is because Lucy wanted to remain anonymous by using a pseudonym.

Recently, books have been published that detail various ghosts and apparitions, one of which is *Psychic Quest,* a paperback by Natalie Osborne-Thomason, from Northampton in the UK, who draws on a lifetime of experience and years of research in the field of ghosts. She spoke to me of her experiences and findings, which you can hear for yourself.[8.5] Natalie said that it is now generally agreed that haunting ghosts are the result of the memory of an emotion or trauma becoming locked in the stonework of a building. She referred to this as the mechanism of stone tape-recording.[8.6] As a person sensitive to such things, she is sometimes able to pick up on these.

To me, it appears that the spin-field carrying brain sharing could become locked into the stone's quartz crystals, and subsequently be gradually released. Natalie mentions that ghosts are sometimes seen on the road because of traffic accidents, or on a wartime airfield, where planes have had to crash land. It does, however, seem that all haunting ghosts are linked to memories of emotion and tragedy that are stored in the quartz. It is not the picture of a car or plane crash that is recorded, but the highly charged emotions of the people concerned at the time. These emotions may be felt later by a person in the dream state of trance, and they will thus 'see a ghost'. These stored emotions may well feedback to the quartz, making the whole business self-perpetuating.

When quartz is knocked, it produces a high-voltage spark, as in gas lighters, but Natalie says that when an old building is restored, knocking by the builders can aggravate the ghost; or release the spin fields from the quartz.

On the face of it, the stone recording theory is too bizarre to be considered, but it is based on the evidence of many cases and is the strongly held opinion of many ghost hunters.

It needs to be said that film and TV entertainment depicts an entirely fictitious and incorrect review of ghosts and hauntings, leaving the public with the idea that it's only entertainment, not reality. The evidence from the past and present shows the true reality.

Dowsing

This term refers to two very different phenomena, but they both rely on trance and states of mind where feelings of self are subdued. The first is searching for minerals or water (water divining), and the second is a method of looking for people or objects. With water divining, it seems that the trick is to lose the feeling of self and then rely on a forked stick or wobbly rods to tell where there is water. I believe that what is sensed is a layer of quartz stones being moved by the flow of water, which generates a spin field – an explanation that fits the facts.

When it comes to searching for people or items, again known as dowsing, that same trance state is required, setting aside feelings of self. The dowsing is achieved using a short pendulum, set to swing from one hand at low alpha or theta frequencies, to aid the trance state. Questions are asked of the pendulum, which swings forward and back for yes, and side to side for no. By asking the right questions, a lost person or pet may be found, or even, if it has an emotive link to someone, or a possession.

A published example of this tells the story of a small harp that was stolen from a theatre in San Francisco. Dr Elizabeth Lloyd Mayer had bought the instrument for her daughter. She worked as a researcher and clinician, and was an associate professor of psychology, publishing over forty papers in her field. The police were not able to help, so as a last hope, Elizabeth was persuaded to phone the late Harold McCoy, the president of the American Society of Dowsers, who lived two thousand miles away. During their call, Harold was able to confirm, by dousing, that the harp was still in San Francisco, but he needed a map to pinpoint the exact location of the thief, with whom Harold had made a psychic connection. On receiving one, he was then able to pinpoint the location, and Elizabeth subsequently retrieved the harp.

Elizabeth realised that the experience had to change the way she saw the world. She actually said, *'This changes everything'* and subsequently embarked on a study of the paranormal, writing a book, *Extraordinary Knowing.*[8.7] But without an understanding of brain sharing, she was unable to determine how Harold had been able to find the harp. He said his technique was 'direct knowing', which Elizabeth tried to relate to ESP, the sixth sense, clairvoyance, and remote viewing, as all appear to be driven by the same 'unknown mechanism'. However, she had no success.

A similar system of direct knowing is called remote viewing, which was studied for military espionage in Russia, and then in the US in 1978. Some psychics can get the feeling of a remote object, rather than a view of it, by brain sharing with people who know an area.

Two years ago, I had a revealing conversation with Leroy Ball, the then president of the American Society of Dowsers. He claims that he has located three thousand water wells, and also missing people and lost dogs, using what he calls 'focused meditation' with map dowsing on Google Maps, dowsing directly from the screen. He says some dowsers can map dowse using only a paperclip on a light string, the pendulum being used purely as an aid to set consciousness aside, which is the same thing as a medium using his guide.

Spiritualism

Many bereaved people attend spiritualist churches in the hope of receiving a message from their loved ones, believing they still live in the 'spirit world'. Real evidence for the spirit world is hard to find. Could it be that mediums obtain information from the sitter by normal brain-sharing?

You do not have to look very far for evidence of real mediumship and contact made with the 'spirit world', where detailed personal information, unknown to anyone else, is given to a bereaved person. The person concerned is thus totally convinced in the realism of the spirit world and that it proves human survival; even the mediums themselves become fully convinced that they are communicating with dead people. In consequence, there are about three hundred spiritualist churches in the UK, working with five associations. In the US, there are twenty-four associations, and very many churches.

My interest started over fifty years ago, when a famous Australian medium visited the UK to speak to members of the College of Psychic Studies in London. I went along and listened with interest to her detailed experiences and the messages that had come through from her spirit guide. She gave account after account of private details concerning her sitters that she had received from the spirit world. All the time she was speaking, I was thinking to myself that the details she revealed must have been obtained by telepathy. Then, during question time, I naïvely asked *'Were not all the details obtained by telepathy?'* There was a stunned silence, with mutterings and glances on the stage, followed by the chairperson saying in a loud voice, *'Next question, please.'* Clearly, I had made a big mistake. Since that time, I have been sceptical of a spirit world and still believe the medium's information is obtained by brain-sharing with the sitter.

The late Colin Fry was a talented UK medium who gave a series of sittings with TV studio audiences entitled the *6ixth Sense*. The programmes became very popular in many countries around the world. After making mental contact with a bereaved person, he would say *'I see a child named Jack or James'* and would point to the relevant section of the audience, saying perhaps, *'Over there at the back'*. I'm sure he was able to do this because he was visualising the view of himself seen by the bereaved person, which he received via telepathy. This would help him find the correct person in the audience in order to pass on messages from the deceased person. Interestingly, I met him back in 2009 and he mentioned to me that he had always been telepathic with family and friends, but I believe he was quite convinced his readings were from the 'spirit world'. The question I ask is: how much might that have been brain sharing?

Mediums claim that they are only the channel of information from an imaginary spirit guide. The point is, by submitting to his imaginary guide the medium can remove his feelings of self and reach the triple peaks while speaking to the sitter, which is a difficult mental juggling act, even for a psychic. There are two other things that link the medium to the sitter, increasing the likelihood of brain sharing. The sitter listens to the medium intently and, therefore, loses his sense of self, as in hypnosis; and secondly both the medium and the sitter share the common powerful emotion of bereavement, bringing about a sense of rapport between them and aiding the possibility of brain sharing.

I have been made very aware of the annoyance and pain suffered by those in grief who have had convincing evidence of the 'other side' from a medium. I give my apologies for any sadness caused by my alternative view of the evidence.

At this point, I would like to tell the story of Phoebe Payne, an only child who came from the slums of Victorian London. She had always been psychic, becoming a medium at the age of 20. By the time she was 35, which was in the mid-1920s, she was employed by the College of Psychic Studies to find talented mediums (this was the place where I had the encounter with the Australian medium). She discovered that when she was working with a medium, she could affect the messages received from the 'other side'. If she imagined events clearly in her mind, the medium would pick up on them and relay them, believing them to be spirit messages.

Phoebe decided to set up a test with the séance group she was working with. She announced that she would concentrate on details of an imaginary dead fiancé. The spiritualist medium picked up on these exact details, which caused considerable surprise in the group. When news of the event reached the ears of the director, he warned her never do such a thing again or she would have to leave. So, she left.

The significance of this episode has recently been raised in the SPR Journal[8.8], suggesting the tests should be repeated, but as yet nobody has dared rock the boat by repeating the test. Realistically, we must ask ourselves how much the medium's messages received from the 'other side' are in fact only brain sharing?

As mentioned previously, Max Cade investigated several well-known mediums. He discovered that when the mediums started their reading the normal alpha brain rhythms were observed, followed by the triple peaks. Later in the reading the medium's arousal went from low to high, as noted on an ESR, indicating hyper-arousal in deep trance. It's a very special state of mind; one that mediums can achieve at will. This state of mind is the key to spiritualism, being brain-sharing in deep trance plus arousal of the conscious mind, enabling a normal mental analysis and conversation.

In order to understand what they call 'the power of the spirit' or 'spirit realms', I spent two days at a local college of spiritualism in training for mediumship. I found the induction procedure to be precisely the same as I had been taught previously when training for meditation and healing. While at the college, I realised I would never be able to produce the effect to order, which is the essence of spiritualism and healing. In my view that is why most spiritualist churches feature 'trained' but ineffective spiritualists; it is too difficult for the likes of you and me to achieve the necessary state of mind, even though hundreds of people attend the college for 'training'.

Psychical researchers have always tried to understand and put trust in life after death, but the researcher GNM Tyrrell said *'The cult of popular spiritualism, which is our worst enemy, unfortunately serves to increase the misunderstanding and mistrust'*. This comment was necessary because some psychical researchers are spiritualists trying to understand the evidence they observe, whereas that evidence is no more than brain-sharing rather than a misguided belief in the spirit world.

Oneness

I have a book entitled *One Mind*, written by Larry Dossey MD. The subtitle is: *How Our Individual Mind is Part of a Greater Consciousness and Why it Matters*. Larry is a writer, broadcaster, and the executive editor of *Explore: The Journal of Science and Healing*. I attended one of his lectures and spoke to him afterwards. While he had no idea **how** one mind occurs, his fascinating three-hundred-page book is filled with examples of it.

One example is when two family members experience similar sensations or physical changes when apart (tele somatic events). An incident Larry quotes concerns a man who went sailing while his wife stayed in bed. Suddenly, the wife was woken by a severe pain in her mouth. Later, her husband returned with a bleeding mouth after being in a sailing accident.[8.9]

Larry quotes several examples like this, and as an identical twin, he has had similar experiences of his own.

Sometimes a person will get a sudden feeling of being stared at. Larry gives examples of soldiers who had an uneasy feeling of being watched, or, conversely, witnessed an enemy's startled reaction when in his sight. Long-range photographers say the same, and policemen are trained not to stare too intently when observing someone because they might realise they are being followed. It also happens in ordinary daily life, for example, when out walking or in a shop. The contributing factor appears to be the intensity of the gaze, bringing on a sense of oneness that can sometimes be felt. It even works over closed-circuit television. It has been found that when someone is being stared at, their nervous arousal (EDA) increases due to an unconscious feeling that their personal space has been violated. The mechanism for the sense of being stared, of course, must be brain sharing.

Larry quotes examples from Rupert Sheldrake's book, *The Sense of Being Stared At*. In his book, Rupert gave the combined results from thirteen thousand, nine hundred tests, where the sense of being stared at was generally recognised. In fact, the odds against chance were astronomically significant, being ten with twenty zeroes after it (more than a quintillion) to one.

Many pairs of identical twins are repeatedly reminded of their oneness. Some identical pairs, separated soon after birth to live with different families, have been shown to lead incredibly parallel lives. Larry makes the point that identical twins brought up together like to show some individuality, whereas twins separated at birth are subconsciously the opposite, having common experiences by brain sharing

Another aspect of oneness is where someone with a developmental disability, such as autism spectrum disorder, demonstrates profound capacities or abilities far more than what would be considered normal (savant syndrome). These people have a skill in one or more of five major areas: arts, music, calendar calculations, arithmetic, or spatial skills. They know or remember things they were never taught; some say due to 'ancestral or genetic memory'. An example of this came from the American Laurence Kim Peek, who had abilities far in excess of the average, but was the inspiration for the main character in the movie *Rain Man*. It seems logical to me that the origin of these mental abilities is brain sharing.

Larry says it may also involve '*that greater meeting place of consciousness*', the *One Mind*.[8.10] Carl Jung understood this, calling it the Collective Unconscious. Larry Dossey talks of entanglement, but no hypothesis has yet shown **how** this happens.

It is only by starting with the phenomenon of instinct, as I did, that a theory may be developed. Instinct is downloaded before birth by the process I call brain sharing, akin to telepathy, using the mechanism that in later life is called the Collective Unconscious or Larry's Greater Consciousness. I have tried to explain **how** this mechanism works – it is triggered by deep trance or lucid dreaming, even though the person concerned may feel awake at the time, but in REM.

As well as the unconscious person-to-person sharing of information that may leak into consciousness, we have mystical experiences, where some of the greater consciousness is downloaded. Sometimes this is the result of prayer, and there are 24 different names for it, each associated with a particular faith. But there are other triggers that lead to the experience, for example: fear, terror, NDEs, speaking in tongues, hypnosis, healing, drug use, Ouija boards, etc., or, in my own case, a combination of various meditation techniques. But what becomes clear is that all the experiences are part of a spectrum of brain and mind sharing, having one common fundamental mechanism: the instinctive one mind of the human species.

In Chapter Seven of Larry's book, *The One Mind of Animals and Humans*[8.11], he details cases of mind sharing between animal and animal, animal to human and human to animal. He says there are thousands of cases in which 'man rescues dog', or 'dog rescues man'. These are incidences where crisis and emotion are shared across species, enabling the rescue. Pets tend to 'know' their master's intention to feed them. Our black cat would be seen walking up the garden path when my wife decided it was feeding time. Another example concerns the American president, Abraham Lincoln. When he was assassinated in 1865, at a theatre in Washington DC, his dog at the White House barked uncontrollably, rushing around in a frenzy. He made it clear to the staff that he knew his owner was in danger.

I like Larry's story of a racing pigeon from Northumberland that went off course during a storm in France, ending up in a pigeon loft in Morocco.[8.12] It went south instead of north, but finished up in a loft on the exact same longitude as its home one in the UK. This is because pigeons are guided home by being aware of the sunrise in their home loft, as well as their natural compass. Later, the pigeon's nephew seemed to know its aunt was at the same longitude and flew to see her – some 1,600 miles away! Once he got to Morocco and was identified by leg tag, he found his aunt's loft by mind sharing with her – problem solved!

I don't find that story so strange, because some years ago I published an article in the *Pigeon Press* detailing how these birds find their way about. It relied

157

on a report explaining that when the lights are left on all night in the home loft, racing pigeons become unable to find their way back home. I concluded that this is because they are unable to gauge sunrise in their home loft, by using brain sharing with their family at home. A pigeon away from home needs to compare sunrise where he finds himself with sunrise at home, to gauge his longitude position relative to the home loft. With that information, he can fly directly home by using the earth's magnetic field to give him latitude. Similarly, at other times, a comparison of sun positions indicating time will allow a bird to find the home loft. This has been demonstrated to be the case, even when the home loft is on a boat that has been deliberately moved.

You might think that there is no scientific interest in all these claims. The fact is there has been much interest in many aspects of the mind, despite little research money. Some of these experiments are now detailed.

In 1989, it was shown at the Mind Science Foundation in San Antonio, Texas, that during hundreds of sessions, the autonomic nervous system could be influenced remotely. The aim was to influence a distant person's EDA by staring at a TV monitor of them, which produced a significant change in EDA compared to periods of non-staring.[8.13]

A programme of research has shown that the understanding of healing is helped by imagery or thinking without words.[8.14] The visualisation of the healer can influence the mind of their subject, to aid the healing, where there is common rapport and deep trance.

Human red blood cells in salt solution tend to deteriorate, but it has been found that a person in another room could, by directing their attention, mentally slow down the deterioration of the cells.[8.15] Measurements were made by shining light through the saline, to obtain a numerical value for the deterioration. Using that same weird control of life, the experimenters were able to remotely control the movement of small knife-fish. Similarly, Matthew Manning[8.16], a UK psychic and healer, influenced the activity of a gerbil, making it slow down on its wheel by using 'goal-directed imagery, focused attention and goal-orientated intentionality'. These big words just mean relaxed concentration.

The paranormal is a confusing subject because there has been no indication as to how or why these effects work: too many effects and too many different names for much the same thing. The same applies to OBEs, which I discussed earlier, noting that they sometimes occur in hospitals in connection with an NDE.

The literature about OBEs is large. They used to be known as astral projection, soul travel or spirit walking. Most people claiming to have had a spontaneous OBE report that they were on the verge of sleep shortly before the experience, or their sleep was not very deep, or they were overtired, stressed or ill. The literature generally says that the OBE was the result of a lucid dream, where the person remembered seeing their body from above, or they travelled away from the bed.

In the previous chapter I mentioned my discussion with a lucid dreamer, called Beverley, who says she has had OBEs as well as mind sharing experiences. She also can heal, indicating a common cause namely trance.

It's claimed that about ten per cent of people have had one OBE experience of some sort; some claim the figure is more than that, some less. It is also said that lucid dreams and OBEs are psi-conducive states, in which a relaxed mind can access the dream state. Many tests and experiments have shown that the effects are rare, although occasionally very real.

OBEs can be induced by drugs, drumming, binaural induction, magnetic stimulation of the brain, sensory deprivation, or even sensory overload. But, overall, the mental state is associated with lucid dreaming, with 'mind awake body asleep', which is about all that can be said from a scientific point of view.

Because the whole thing is so puzzling, there has been no scientific recognition. But there is some acceptance of the idea that we are all fundamentally connected as part of an underlying common consciousness. This book makes the point that there is only one mind – because life is a perpetual copying machine, replicating brain and mind indefinitely.

Sleepwalking

It's a frightening thought that we can do things at night of which we have absolutely no memory. When I was eleven years old, I did something strange and stupid after walking about at night; I somehow got into the wrong bed, between the occupant and the wall. In the morning I was frightened, so perhaps that's the reason it has never happen again, but some people do sleepwalk repeatedly.

You can hear my discussion with Lee Hadwin (podcast no 24) who had unsolicited fame around the world as a sleepwalker. He started by making drawings on the wall at night, then made amazingly good artistic drawings with the pen and paper he left around purposely. What is amazing is that he had no particular training or aptitude and cannot approach his night-time standard during the day. Some of his pictures sold for $4000 in the U.S. and some can be seen on the Internet.

Drawing whilst sleepwalking is not unique as there have been other cases noted, Matthew Manning in the UK, and a man in Australia. It is the quality of the drawing that is so amazing being almost works of art where the talent seems to come from elsewhere. I think it is reasonable to assume that during a night-time trance a psychic connection is made with a good artist.

Speaking In Tongues

Humans may do strange things when in trance, although perhaps the strangest is speaking in tongues (glossolalia). By reverting to unintelligible childlike speech during meditation a deep trance is reached which brings emotion, uplifting satisfaction, and mild euphoria.

The relaxation and regression of mind back to the days of earliest speech, produces unintelligible words and sounds without using the speech area of mind. This is like the repetitive use of a mantra to encourage meditation. The brain goes through the motions of producing speech, without mind becoming involved. Consciousness just accepts what is being said without questioning, looking on as in a dream, and it is because the mind is occupied listening to the words being produced that the trance state continues. Reality testing is inhibited and if the trance of meditation goes deep enough, ecstasy and religious enlightenment is the result.

The practice has always been part of religious services. Ecstatic utterances of a divinely inspired nature are mentioned in the early Egyptian writings.

Plato and Virgil recorded phenomena like glossolalia. Almost all religious movements have, throughout the ages, made similar reports and it is only reasonable to assume that all these highly personal experiences, come from one root cause, and use the same psychological mechanism. There are many references to it in the Bible; Acts 10:46 says *'For they heard them speaking in tongues and praising God'*.

Praying in tongues and healing promotes a sense of camaraderie in the group united by the phenomenon which they experience together, but invariably their unusual behaviour makes them outcasts from the broader society of the church. Unfortunately, this antagonism can too easily split a church, which has given rise to many different sects even within Christianity. These negative comments on the experience, in no way detract from the fact that the euphoric experience leaves the fortunate recipient with a sense of presence and knowledge that he will be affected by, and remember, for the rest of his life.

Now a brief description, with my impressions, of a church service I attended a while ago in Manilla. It is held regularly in a hall, rather than a church, and in common with other churches, the music is supplied by a modern style group -- of guitar, bass, keyboard, and drums. Choir is provided by two ladies with microphones, but the loudspeakers are powerful, with a sound mixer at the rear of the hall. It seats 150 people and is normally full. In an extremely open and friendly way the preacher or leader makes a few opening remarks, followed by a swingy piece of opening music. Towards the end of this, the preacher speaks over the music to announce the first hymn, the words of which are put on the wall by an overhead projector. The musicians carry straight into the hymn, with the words being moved up on the screen so that everyone must look upwards to read the words.

Continuing in this way, with the one hymn leading into the next, for about 20 minutes, all apparently in the same key, the religious fervour increases with the preacher singing and interjecting into the hymns in the style of gospel singers. Then she imperceptibly changes to glossolalia, intermixed with praise and Hallelujah. The music becomes mainly monotone, and the congregation is in a position of being able to join in the tongue prayer, privately, without it being obvious, enjoying the excitement of the event, or even just looking after the children of which there are quite a few. I saw no exhibitionism, nor anyone who was obviously entranced during the five or ten minutes of glossolalia. Gradually the leader lowered the level of excitement and reverted to quiet praise and thanks for receiving the Spirit of the Lord.

A sermon was then given, followed by the usual announcements and final hymn. I found the experience produced a strange mixture of feelings, of being oppressive, but elated and emotive at the same time.

This description was written straight after I attended the church with my sister-in-law who is a devout born-again Christian. As I had recently read a book that was published 50 years ago, giving many graphic descriptions of services in which tongues were used, I had no real surprises except for the way in which the whole thing had been modernised. I had not expected the use of a modern style band with high power amplification, nor the very smooth presentation which ensured that there was absolutely no break at all from the opening remarks until the start of the sermon. Clearly it was organized to be very hypnotic.

But the real eye-opener was the incredibly homely and friendly atmosphere. To become part of it for a while was an uplifting and emotional experience. Most tongue speakers appear to have had an anxiety crisis preceding their first experience due to personal life difficulties. Subsequently, the euphoric feelings subside and generally the speaking of tongues is described as pleasant and uplifting rather than euphoric. Once the ability is possessed a person retains it and can speak with fluency whenever he chooses. It matters not whether he is alone or in a group. He can speak in tongues while driving a car, or swimming. He can do it silently in the midst of a party, or aloud before a large audience. It does not involve conscious thought at all, in fact quite the opposite, but the senses continue to operate during the experience.

Further information is given in the book "The Psychology of Speaking in Tongues" by John P. Kildahl, giving a great deal of background information. His study of the subject, which lasted ten years, from 1962 to 72, was undertaken with research grants in USA. These were given because of the sudden rise of glossolalia in the Episcopal, Lutheran and Presbyterian congregations, with the aim of helping the Church in preparing guidelines for church policy. Many hundreds of people were interviewed in this extensive scientific study. Previously the subject had received little scientific attention -- none the less Kildahl was able to draw on 334 references on the subject -- the list of books and papers is given in his bibliography.

Psychic Interaction

We see a similar state of confusion regarding the possibility of interaction between mind and material things, for example, poltergeist activity or spoon bending. Having never witnessed anything like this myself, I have some scepticism, but on the other hand some of the evidence is hard to ignore.

If you have around forty minutes to spare, watch Uri Geller demonstrating his psychic abilities under close scientific scrutiny at Stanford University.[8.17] It's most fascinating and involves copying pictures and 'seeing' inside sealed envelopes and aluminium cans. He can also 'see' the fall of a dice inside a box, and he deflects a compass and alters the reading on a magnetometer. The test I found most impressive involves a copper ring, which Uri squashes using light finger pressure, when it would usually require 150lbs of weight to deform it.

While some people have denounced Uri Geller as a conjuror, they have not attempted to repeat the same feats, even though James Randi pretended to repeat a few feats by simple conjuring rather than psychic interaction. Other individuals have shown psychic interaction with matter. Nina Kulagina was a Russian psychic who joined the Red Army tank regiment at the age of fourteen where she learned to mix the trance state of fear with hyper-arousal. She could move small items, such as a matchstick or compass needle, by psychokinesis.

Spiritualists have reported various psychic phenomena, such as table tilting, producing objects (apports) and lights (orbs). Poltergeists have been reported as having moved objects, raised fire, or produced water. But any credibility of psychic interaction has been destroyed by cheating, and by highly misleading TV entertainment.

Reincarnation

There have always been reports of cases where the system of inheritance has gone wrong, where children are born resembling someone other than their own family. This resemblance can be physical, but there are also memories of events or personal details. It is known as the transmigration of souls, or, more usually, reincarnation. My dictionary, (Collins English dictionary) describes reincarnation as the '*Rebirth of a soul in successive bodies*', which is the general understanding, but I don't believe this is the correct interpretation of the evidence.

It is often suggested that someone might be an ancient emperor reborn, or that next time he might be a mouse or a worm. This is incorrect, as it is not part of the unexplained evidence, which is sometimes called the World's Greatest Mystery. Many religious faiths accept reincarnation, but details of the evidence are so bizarre that most people regard the idea as impossible and make-believe.

Dr Ian Stevenson, who was a professor of psychiatry at the University of Virginia, became internationally recognised for his work in scientifically quantifying the evidence for reincarnation. He travelled the world and wrote fourteen books and three hundred papers on the subject. The book I have[8.18] *Where Reincarnation and Biology Intersect* is a summary of his two-volume treatise comprising more than two thousand pages. His team collected details of about three thousand cases of children who appeared to have had a past life. They remembered many details of another person, which led to the belief that they were that person born again, sometimes a member of the same family, sometimes not.

Similarities also included birthmarks, deformities, unusual abilities, and phobias, for example, as well as memories from the reincarnated person, including their names and where they lived. Birthmarks and defects, which corresponded to wounds on a deceased person, were noted on a child in half of Ian Stevenson's cases. Strangely and significantly, these were nearly always associated with cases of violent death.[8.19] My hypothesis is that when the relative's emotion causes trance leading to brain sharing with a growing foetus, it can continue for a few years.

The evidence for reincarnation is not only from birthmarks on a baby and its verbal descriptions of unknown people as soon as talking begins, but from behaviour that is associated with the relived person rather than what has been learned from the child's family.[8.20] The strange behaviour can also be a phobia, a desire for different food or clothing, or an addictive substance. A child will speak of the unknown people as soon as talking begins, sometimes referring to another mummy, brothers or sisters. This will be between the ages of two and eight years old, after which time the memories seem to fade due, I believe, to the fact the brain-sharing connection is lost.

As part of his analysis of the evidence, Stevenson gives four possible rational explanations for reincarnation. Firstly, ESP; secondly, possession; thirdly, that the child's mother knew of the deceased person's wounds; and fourthly, reincarnation. He is left undecided, but he does try and expand on the idea of ESP by suggesting that it is a system of mind-carrying[8.21] (psychosphore). That sounds the same as what I refer to as brain sharing, but he is at a loss to find any explanatory mechanism, as he does not accept brain sharing.

So, what can be said about explaining the evidence of reincarnation? We are aware that an embryo grows initially in the form of its species, but towards the end of its growth period the information of inheritance will determine its features and behaviour from its parents, with whom it shares its DNA. My hypothesis is that the fetus might begin to receive inheritance information by brain sharing with a stranger **who has similar DNA**, where the trigger is anguish and traumatic emotion in that person. Traumatic emotion is likely when there has been a violent death, which explains why fifty per cent of reincarnation cases are associated with this type of sudden end to life, sometimes producing deformities or birthmarks in the fetus. In these cases, the anguish is felt by the relatives of the deceased person, who then becomes linked to the fetus by brain sharing over a period of several years.

In the cases where violent death does not appear to be relevant, the link by brain sharing could be triggered by some high intensity emotion in the person that is subsequently reincarnated, which could be anybody from a relative to a person living far away and unknown to the family, but whose DNA is similar to the child's. That link is then continued for perhaps a few years, during which time the child receives personal details of the stranger or relative, including even names and addresses. The child just seems to know, as if it has memories of being to the other house.

I suggest that these scenarios fit the facts of reincarnation, and the mystery can thus be viewed as something extraordinary but understandable. The mechanism is inheritance that has been misdirected from a person other than the parents, due to similar DNA. It seems that at a particular stage in the formation of the fetus's brain, when instinctive information is being recorded, a connection is made with someone who is troubled with deep emotion. That connection, which lasts a few years, appears to sometimes alter physical features in the fetus, such as adding birthmarks, but after delivery telepathic information is still received, resulting in a bond of knowledge similar to that of twins who are separated at birth. If the strange facts of reincarnation are regarded as an example of telepathy, where all minds are one, then there is no more mystery.

Another mystery that has baffled scientists concerns heart transplants. A human heart contains ten thousand neurons (brain cells), so it is often called the 'little brain'. When a heart has been transplanted, memories are occasionally transferred across to its new owner, for example, preferences, habits and behaviours. It seems unlikely that a heart would normally contain behavioural information and other memories, but its DNA might become accepted by the brain of its new owner. Consequently, the new owner would link with the donor's family, initiated via the donor's barcode and by brain sharing, and that would subsequently enable details of preferences, habits, and other memories of the heart donor to be transferred across.[8.22]

In all these cases of reincarnation, the explanation is fairly obvious when one considers my new core theory, showing how inheritance is **directly** transmitted from parents to the fetus. The answer is that the child's DNA, which is half from each parent, is used to identify their details. But the fetus sometimes starts to accept inheritance information from someone with similar

DNA, especially where there is a high level of emotion. This is because a fetus looks for emotional information from its parents, which gives an explanation for the World's Greatest Mystery.

Making Sense of Life

9. Evolution and Inheritance

Current Doctrine

Today, we have an accepted mechanism that explains how all the species of the world have evolved from a common ancestor by slow changes in inheritance. Animals and plants adopt the required change very slowly, due to random changes in their DNA caused by mutations. If the change is an improvement in the fight for survival, then offspring with this trait are more likely to survive. We are led to believe that this is the way creatures have evolved, but I believe mutations are no answer to the problems of evolution.

My Collins English dictionary says that 'doctrine' means *'stubbornly insistent on the application of a theory without regard to practicality'*, which is a good description of the above doctrine and the arguments that have raged for well over two hundred years.

Fundamental Difficulties

There is a wonderful book *The Living Stream* by the Oxford zoologist, Sir Alister Hardy, about the living stream of evolution. He likens the flow of evolution to the flow of a river and quotes Dr Erasmus Darwin, who, in 1796, said, '*all animals undergo perpetual transformations, which are in part produced by their **desires** and aversions . . . and many of these acquired forms or propensities are transmitted to their posterity.*'[9.1] Today, we say inherited rather than transmitted, but we say it quickly in the hope that no one will ask **how**, because there is no answer.

Charles Darwin, the grandson of Erasmus, went to great lengths to give evidence in support of his grandfather's theory, but neither of them could make up their minds over the question of **how** inheritance worked: was it mutation or inheritance of acquired characteristics? Neither idea gave any explanation as to **how** it might work or was 'transmitted'. Perhaps Erasmus Darwin wisely understood that all life is due to **direct** 'transmission' from parents to offspring.

This is the fundamental missing link in all theories of evolution, which persisted until about 1925, when the suggestion was that a Modern Synthesis of many opposing theories was the answer, based on observations of genes and chromosomes. But again, it did not explain **how**, except that it was suggested that mutations initiated the necessary changes.

When the structure of DNA was revealed, it was thought that the issue was finalised; Lamarck was definitely wrong, there could be no inheritance of acquired characteristics and it was all down to DNA – **but how?** Today we have Neo-Darwinism, formerly the Modern Synthesis, which relies on the idea that DNA is the book of instructions for life, where mutations lead to changes of inheritance that evolve into new forms of life. To suggest that the evolution of eyesight, consciousness and the immune system are the result of random mutations is clearly ridiculous, but scientists don't agree.

It seems that the inheritance of acquired characteristics and habits might well be the driving force of evolution being 'a matter of common sense', according to Charles Darwin. Neither Lamarck, Erasmus nor Charles Darwin could indicate a mechanism for inheritance, but we only have to look at the features of our own family to see the effect.

Evolution

When at school, my Christian education did not allow any mention of the dreaded word evolution. It was left to a Jehovah's Witness friend and colleague, who once while talking to me about religion, embarrassed me into an admission that I knew nothing of Darwin or evolution. He promptly lent me a little book to make his point, but within an hour, much to his dismay, I was a convinced Darwinian. I have no difficulty in the belief that the Lord created heaven and earth, but in his own time, rather than the seven days written in the Bible. It is the theologians who have such difficulties, rather than their congregations.

The method Darwin used to promote his theory of evolution in *The Origin of Species* was to show that each species changes very slowly into varieties that are best suited to the available food, environment, and predators, which he called 'variation under nature'. Some varieties changed, others did not and some became extinct. To make his point he compared 'variation under nature' with 'variation under domestication' which happens very fast.

For thousands of years, we have improved our food by constantly breeding from the best natural foods, like rice and wheat, and we have created new varieties or sheep, cattle and dogs. Originally, this was perhaps almost unconscious selection of the best ear of corn or the most useful from a litter of dogs, although it was then realised that conscious selection of characteristics was worthwhile which increased the speed of change.

Darwin noted that the features of Pointer dogs had greatly changed in 100 years. Similarly, the Rock Dove had been changed into many varieties, from racing pigeons to grotesque fancy pigeons with tails like fans (fantails) or the pouter with a hilariously large chest. Today we keep a variety of wolf for a pet, that has been altered to a Chihuahua, which would have no hope of survival in the wild.

All this makes the point that, in nature species and varieties can change and have been the cause of evolution over 2000 million years, all the time being subject to natural selection and survival of the fittest.

In the first edition of Darwin's revolutionary book, he relied on acquired mutations to alter inheritance but, eight years later, a professor of engineering pointed out that the idea was flawed because any mutation would be blended away after a few generations. Darwin was devastated, writing to Wallace, the man who also discovered the theory of evolution, that he was blind and had been deceived into thinking that natural inheritance acted in the same way as man's selection of varieties.[9.2] Blending inheritance reduces a mutation change to three per cent after five generations, but with continued selective breeding, ninety-six per cent of a change will be retained. In later editions of his book, Darwin relied more on Lamarckism which again did not indicate **how**, and so today we still have that same fundamental difficulty: how does inheritance work?

Another problem with Darwin's theory is the fact that neutered insects in social colonies cannot hand on their elaborate instincts by DNA, which poses an unanswerable question. Similarly, the instinctive behaviour of some insects after laying their eggs cannot be explained.[9.3] How can an amoeba without any sense organs or eyes build itself a house out of bits and pieces from the seabed? [9.4] It is only by giving a new insight into the mechanism of inheritance that these problems will be resolved.

The Nature of Inheritance

Sir Alister Hardy gave lectures on natural history and religion, in which he gave his thoughts on inheritance, intelligent design, evolution and instinct.

Hardy noted that Lamarck had stressed how inheritance referred to changes of habit brought about by changes in the environment. But at the same time, Lamarck had suggested that the long neck of a giraffe was due to wishes or **desires**. Hardy thought the word 'need' would be more appropriate[9.5] as he could see no way that a **desire** could be encoded into DNA.

Sir Alister then quoted an important milestone in the understanding of inheritance, which introduced the idea of a common subconscious in insects to explain the mysteries of instinctive behaviour. The suggestion was that the web spinning instincts of spiders may be due to the individual spider being linked up to a larger system, in which the web spinning experience of the species is stored.[9.6] This suggestion had been made over seventy years earlier in the book *Telepathy* by Whately Carington.

Sir Alister then related this idea to the fundamental difficulties in biology (mentioned previously), and how it might offer a *'highly speculative alternative to current evolutionary theory'*. He went on to describe his speculative idea that not only may the instinctive web spinning of the spider be subconsciously shared, but that all the details of its form be similarly shared by *'a sort of psychic blueprint'*. He said the living stream of the species – the psychic stream of shared form and behaviour ... *would flow on in time parallel to the flow of the physical DNA material.*[9.7]

This was a huge and very bold step. But there still remains that one difficulty: **how** does the psychic stream of information relate to DNA? **How** do the inheritance changes become encoded into the parent's DNA, so that they may be transmitted to the next generation? This is the same problem that Erasmus Darwin had encountered in 1796.

And the answer, I suggest, is that the parent's DNA does **not** get encoded, because there is no way this could happen. The changes do not have to be encoded because the building instructions of the child do not come from its DNA, but are copied over, via cell-to-cell transfer, by inheritance from its parents. So here we have a clear understanding of **how** acquired characteristics are inherited, which now answers the question raised by Erasmus Darwin and Lamarck.

The DNA of a child is only a randomly shuffled combination of its parents' DNA, producing similar features, behaviour, and habits, which we call inheritance. I believe this happens automatically, as all the information for growth of the body, brain and mind is **directly** copied over from the parents. The looks, behaviour, intelligence, and the acquired characteristics of both parents, will in this way be passed on to the child along with any strongly felt unconscious **desires** (of either parent). Life is an eternal copying machine.

With this explanation, the problem of inheritance is neatly solved: DNA directs life because it is a digital identity code, but it does not provide the building instructions. The form of the body, physical features and the details of the brain and its control system are copied over from the parents, being unique to that individual and its DNA. The DNA identity code in every fertilised cell that survives is totally unique, so it will be associated with behaviour and appearance that is exclusive to the child subsequently born. It is apparent that this change in every generation, known as diversity, enables a gradual change in the form of every subsequent generation, leading to alterations in inheritance, evolution and the resulting form of the species.

This new hypothesis reveals that doctrines that attribute the building instructions to DNA, as well as the belief that mutations lead to evolution, are without foundation. A new theory of inheritance and diversity, which shows that the lifestyle, behaviour, and the unconscious desires of parents are passed on to the child during its growth before birth, allows a new understanding of the evolution of each and every species.

Interbreeding

It seems to me that there is yet another factor in evolution, which is interbreeding, leading to a new species. Interbreeding in nature or by design is common in plants and fungi, as well as in mammals. The mating of a zebra and a donkey, known as a zedonk, or a zebonk, can have a donkey body with striped legs. Similarly, a zebra/horse cross, known as a zorse, can look like a horse with a striped back or a fully striped head. This type of mating generally produces semi-sterile stock, although back crossing (mating with relatives) usually increases their fertility.

The general reaction to interbreeding is that offspring are sterile, but this is not true. Even a mule, regarded as always sterile, has been found (by the Chinese) to occasionally breed successfully.

Another example is a fish-bird cross, which was thought to be the result of the accidental transfer of sperm from a wagtail bird breeding by the sea to fish eggs in the water, which resulted in fish with the heads of birds.[9.8] Several of these have grown to full size and then been caught and photographed, showing that this is a rather unfortunate possibility. It makes the point that interbreeding between the most unlikely species is possible, and this is probably another factor in the evolution of new species.

When it comes to the definition of a species, it's difficult to be exact. Perhaps the best that can be said is that: *A species breeds true and survives for many generations in a particular location, where its varieties are all subject to survival of the fittest and natural selection.*

We don't now have to think of a species as being fixed by its unalterable DNA; it's only where interbreeding, or the other form of evolution, produces a successful variety that we should use the word species.

There are thousands of hybrids produced naturally, or by man – plant, fungi or mammal – that only make a new 'species' when they have successfully overcome natural selection. An example is a breed of pig-sheep cross that has a woolly coat to help it survive the cold winter. In Hungary, this animal is known as a Mangalica.

Then there is a cat-rabbit cross, known as a cabbit, a chicken-pigeon cross, a cow-horse cross, and a leopard-lion cross called a leopon. All these crosses may be semi sterile, which could be improved by back crossing, as I have already suggested.

A major study of interbreeding, or hybridisation, was carried out in the US by the geneticist E. McCarthy PhD. He pointed out that hybrids have been a big factor in the story of evolution, making the Darwinian evolution theory incomplete. He also said that we are **not** directly descended from monkeys.

He hypothesised that two hundred million years ago, there was a pig-bonobo cross that produced a monkey with some of the internal and external characteristics of a pig. It walked upright without a hairy body, but it developed the bonobo's chatter into speech, and then learned to think in words. This successful hybrid now dominates the world. McCarthy said about this hypothesis: *'For the present there's no need to make a definite decision on the matter, but certain lines of reasoning do suggest the idea should be taken seriously'.* I take the evidence and lines of reasoning he presented seriously but see his website [9.9] and decide for yourself.

As a geneticist, McCarthy has presented his evidence and hypothesis in the belief that DNA is a book of instructions, showing the way that different DNAs and even numbers of chromosomes belonging to the two parents somehow combine, to directly form an amalgamation of the parents' physical characteristics. I do not think the proposed amalgamation of the chemistry of the two DNAs directly controls life, and I suggest that my alternative theory that DNA information indirectly forms life is a more likely idea. It seems highly unlikely that the DNAs of the fish-bird cross could successfully breed, but it produces peculiar individuals in the same way as all other interbreeding occurs; and they would be sterile because a cross between very different species is almost always unable to reproduce.

Evolution in Action

Inheritance, diversity, and the survival of the fittest slowly changes the DNA of every species over time. But it does seem that unconscious **desires** are inherited and alter the course of evolution. So, Lamarck was right after all. Having said that, is there any evidence?

Some butterflies have exactly mimicked the wing patterns of poisonous butterflies; others have made themselves look like a leaf – all by **desire**. Birds of paradise have altered their plumage to please the **desires** of females. To suggest this was all done by mutation is ridiculous. This was obvious to Darwin when he went to the Galapagos and saw how finches had altered the shape of their beaks to suit the type of food they ate, but he was afraid to use the word **desire**.

When it came to the evolution of the human eye, Darwin could find no possible explanation, saying it gave him a 'cold shudder', but surely gradual modification to produce better vision by unconscious **desire** would explain it? It seems to me that an important factor in the evolution of all animals is the unconscious needs and **desires** passed across by inheritance. This looks like, and is in fact, the work of intelligent design, guided evolution, or purposive intelligence. Even flowers and trees are influenced by the needs and desires of animals.

In Russia, sixty years ago, Dmitry Belyayev started selecting farmed silver foxes for tameness – they had previously been fearful and aggressive. In four generations and only ten years, some were reasonably tame. The tests continued, and now eighty-five per cent are as tame as dogs, although some still possess the wild trait. Strangely, while the selection was done only for tameness, the foxes have changed their fur colour and physical characteristics to be rather like dogs, as well as being as friendly as them.

In a similar way, human consciousness seems to have advanced quickly, because of an unconscious **desire** to develop speech and use tools. I found an example of this when talking to an acquaintance, who showed me his strange double-jointed thumbs. He said he thought it had happened because his mother had fallen and dislocated both thumbs, making his thumbs double jointed by the mechanism of inheritance. This could not have happened by the mother's DNA changing after she fell – it could only have occurred because her experience had directly affected the growth of her child due to an unconscious **desire** that thumb joints can bend backwards. This is a good example of acquired, and unconsciously desired, characteristics.

I am amazed by the evolution of frogs and toads. In the UK they mate and lay spawn in ponds, from which the young tadpoles emerge a few weeks later. But in the tropics, where there are many predators, some male frogs rear the tadpoles in a mouth pouch, while others have pouches in their sides, from which the baby frogs emerge. In desert conditions, one type of female toad burrows down and rears her family in a pool of foam she has made, while another female is so large that the males develop a sticky chest in order to glue themselves on at mating time, which is sometimes followed by a procession of hopeful males all stuck together! The question is, how is it that these peculiarities of habit and physical form have evolved? Surely it can only be their unconscious **desires,** brought about by a change in climate or predators.

This new perspective shows **how** and **why** evolution generally results from the process of inheritance and diversity. It comes about thanks to my revolutionary view of the function of the DNA digital identifying code. The minutest thousandth part of our DNA, which is randomly produced by the shuffling of our parental genes, is personal to each one of us. Our personal DNA denotes the passwords of our parents and grandparent's lifestyle, experiences and, as I have stated before, their unconscious **desires**. This is then passed on to our children, together with family likenesses such as hair colour and facial features. The rest of our DNA, created by the random shuffling of parental genes over millennia, our junk DNA, ensures that we grow up to be human rather than a monkey.

The fossil record shows that most species have been stable over millions of years, but there have been sudden unaccountable changes. It is clear that these changes have come about by interbreeding, where the new progeny found a successful niche in which to survive as a new species.

Ever since the first single cells evolved, this same system of using a chemical barcode has been necessary to ensure the continuation and diversity of the species. The fact that all life uses RNA or DNA for its survival does not mean that it is the architect of life, as was envisaged seventy-five years ago. DNA is a long string of essential passwords going back millions of years, but changing each time an egg cell is fertilised, which I am convinced is the main driving force behind diversity, inheritance, and evolution.

10. Discussion

The day will come when this will be given as a curious illustration of the blindness of preconceived opinion.

Charles Darwin

Here, I discuss the way in which preconceived opinion has blinded the study of the science of life. While accepting that problems and anomalies do exist, scientists have found there is no conceivable way to solve some baffling anomalies. Doubt has been cast on the theory that all growth and information flows from DNA, although preconceived opinion has ruled out the possibility that life and inheritance proceeds by replication. Knowledge of the mechanism of inheritance, would have filled the gap in evolutionary theory, as well as give understanding to reincarnation. But blindness prevails where anomalies have been quietly put to one side, mounting up for 100 years, waiting for someone else to find the answer.

Biology

This deplorable situation may be resolved if life is seen as a replicating system, where each new generation is a copy of the two parents' families, and arrived at by comparing the personal codes (DNA) received from family members. In other words, life must be seen as a perpetual copying machine.

The basic theory of biology was fundamentally flawed by scientists misreading the evidence, in 1928, with the identification of a transferring principle[10.1] by Fred Griffiths who injected mice with different strains of streptococcus to see which strains killed the mice. This principle was subsequently 'proved', by Avery in 1944, by showing it was the DNA molecule transferring intelligence of a particular strain. Avery's 'proof' was obtained by using a process of elimination in Petri dishes rather than injecting mice. However, it is now possible to prove that it was not the movement of DNA material transferring the growth information, but in fact only the DNA code enabling the transfer of genetic information from one body to another by a mechanism unimagined by science.

The premise of the transferring principle hypothesis was mistaken in 1928, leading to the false idea that DNA is an instruction book, although that preconceived opinion has been falsely taken as the fundamental principle of biology ever since.

I have shown how reincarnation information is **transferred** to a child before birth, based on the close similarities in DNAs, that allow the normal process of inheritance. I note that this **transfer** of information, copying physical and mental information from a person to a fetus is not the **transfer** of DNA molecules, but only the **identification** of the DNA code. This is the **proof**, indicating that Avery's results were misinterpreted in 1944, leading to the absurd situation that biologists face today, where dogma incorrectly states DNA particles to be the instructions for life. The correct interpretation of the tests has been missing for 93 years, so it should now be possible for scientists to rethink genetics, and all the anomalies which flow from their mistaken beliefs. Only very recently have I realised the significance of seeing the 1944 transferring principle in a new light, which reveals how and why biologists have been blinded by preconceived opinion.

We are told by the scientists that inheritance, which we take for granted, is transferred across with our parent's DNAs. But when it comes to finding a mechanism, by which DNA records and transfers the information, scientists have a problem which has been solved by several different hypotheses. In fact, five different methods have been put forward,[10.2] merely noting the natural chemistry or processes of life, none of which are sufficiently acceptable to have been generally agreed. It is only if inheritance is viewed as a copying process of body and mind, that the mechanism is perfectly obvious. It has been the lack of any method, by which inheritance of parent's characteristics have been passed on to their children, that has frustrated efforts to further understand evolution.

Now that we can understand Lamarck's 'inheritance of acquired characters', we can immediately see the 160-year-old dilemma of evolution is solved. Furthermore, it is clear how acquired and desired characters are inherited, which I believe Lamarck had suggested. He used the word 'besoin' which in French means need or desire, leading to the confusing situation that scientists of evolution have argued about for years. Did he really mean desire?

The old theory of DNA could not explain some awkward evidence in genetics, (trans-generational inheritance). To explain that evidence, the unlikely hypothesis of epigenetics was put forward. A paper, entitled *Epigenetics - It's not just the genes that make us,* written for the British Society of Cell Biology by Dr Ian Cowell, explains the situation well without the usual jargon. As I have said previously, I see there is no real mechanism, only observations of cell change during life processes. The proposed doubtful mechanisms are attempts to explain that difficult evidence, but the mechanism becomes obvious when my proposed theory of inheritance is realized. Dr Cowell asks, how life's epigenetic information in DNA is duplicated when cells divide? concluding it is **'less well understood',** perhaps because life does not work that way.

When it comes to the number of genes noted in each species, there have been difficulties and discrepancies. It was said, we humans should have the largest number of genes due to our super intelligence. It turns out that we have 25,000 genes; but rice has 38,000 causing much scientific difficulty. Also, two varieties of single celled paramecium have very different gene counts. How could this be? For me, genes are a record of historical varieties in a species, so it is to be expected that rice, which has experienced many changed environments over time, would have many more genes than ourselves. The two varieties of paramecium probably lived far apart, where one had a stable environment, while the other suffered many changes leading to an increased gene count.

It is often said that extinct species can be brought back to life; known as de-extinction. This has been done by cloning from a tissue culture of a near extinct species, but de-extinction has not been done from the instructions from old DNA. It is thought that ancient DNA from a Woolly Mammoth could revive the species. This has not been done; perhaps because the idea is incorrect.

Scientists have sent a vehicle 300 million miles to understand why the surface of Mars is the way it is, but they cannot tell me how Ella, my great granddaughter, grew from a single cell to be born with body, mind and an operating system. This is the enigma of life requiring a transmission system (quantum biology) and some way that cells can speak to cells both within a body, and from body to body. What has been discovered is that every living cell, from a bacterium to the cells in our fingers, have hundreds of strange microscopic tubes called microtubules. They have been called biological waveguides, meaning they transmit information along themselves, in the same way as nerves of our body and brain; but they also transmit and receive signals within a body, like the way an aerial transmits and receives radio messages. Science has found a mechanism[10.3] by which the microtubules and primary cilia can both transmit and receive information, but I maintain they also transmit from body-to-body, and mind-to-mind, that accounts for morphic fields, which have been debated for over 100 years. It is known that a human embryo has brain cells by four weeks when it is only ¼ inch long, so it could receive the growth information, transmitting it via nerve cells to stem cells, which then organise the way local cells split to create the necessary growth and form.

My Ella, as a fetus at six weeks old, had a heart rate of 110 beats per minute, her skeleton of cartilage was complete, she could co-ordinate voluntary movement, but most importantly she had recordable brain waves,[10.4] showing that she was in connection with the outside world. Up to that time she was growing like her placenta, as part of her mother's body, but at only six weeks she was taking over and receiving growth information on her own, which is incredible and amazing, as she weighed only 0.2 g and was 6 mm in length.

Before my daughter was born her mother suffered morning sickness but our doctor refused to prescribe the new drug thalidomide. The way this drug worked was little understood at the time, but it resulted in miscarriages or deformities of 10,000 children when taken between the fourth to eighth week of pregnancy. Thalidomide, I think, blocks cell microtubules, so the initial growth information ceases at a critical time, inhibiting the transfer of growth information of limbs in an embryo which is only 6 mm in length at the time. This situation is only understood when the action of microtubules is realized. Alcohol is another gas that blocks microtubules. It reduces sensation and clouds the mind, but to a lesser degree than chloroform.

All my conclusions are based on evidence and scientific observations that have been around for years, giving credibility to my hypothesis. These ideas suggest a new theory of biology, but my critics have suggested they are merely 'unprovable assertions'. They are correct, of course. It's up to biologists to decide whether DNA is just a book of instructions, or, alternatively, whether life is merely a perpetual copying machine, always striving to improve itself.

The process of cell differentiation has been the main challenge for biologists. How does a stem cell organise the correct cells to be in the right place to grow appropriately? It's only if the instructions to copy the previous generation are received, that the correct differentiation is possible, and it's only by understanding microtubules, primary cilia and stem cells, that an explanation can be given.

Scientists, when looking for a new way of understanding information flow in growing cells, have observed what they call bioelectricity between cells. I believe this to be the result of microtubule action, the same as we notice from a thinking brain. This questionable idea has been promoted saying that, by understanding bioelectricity, we will defeat aging and cure cancer etc. Similarly, the ideas of xenobiology in which DNA and RNA become XNA, is another vain attempt to find the missing link in biological theories. Biologists, while becoming increasingly desperate to find the solution to DNA, are searching every possible approach; except what to me seems obvious. The fact, that microtubules are at the heart of all life, appears to have been completely overlooked, as has the possibility of a quantum carrier of biological information.

Quantum Biology

The study of psi always leads to the question: how does it work? Could there be a carrier, perhaps something like radio waves? When Uri Geller first became famous in the UK, a Professor John Taylor witnessed Uri and believed psi was genuine. He argued that the explanation could be found in electromagnetism, but he found no evidence. Five years later, and after many tests, he wrote a typical scientific rebuttal; *We have searched for the supernatural and not found it. In the main, only poor experimentation, shoddy theory and human gullibility have been encountered.* These were the words of a frustrated scientist, who had to put the paranormal behind him in order to get on with his life.

I have tried to show that the evidence for psi is reliable, but seldom repeatable, though the problem is **how** could it work? As the carrier is not in any way electromagnetic and must carry psi as well as biological data, it is probably something unknown to science. I found the answer came from a very unlikely source - the Stargate programme. A paper by James Spottiswoode and the program Technical Director, Edwin May (who I spoke with) gave the amazing evidence of the carrier of telepathy originating in the sky, peaking at 13.30 LST, resulting from an analysis of 2482 successful telepathy tests. It's important to note here, that on its own Spottiswood's report is amazing even if not very revealing, but the reliable reporting of scientific evidence of 2482 positive tests, is in itself conclusive evidence for psi.

I was already aware of the work of Michel Gauquelin who, by comparing the birth times of 35.907 people with their profession in later life,[10.5] had discovered a highly significant relationship between time of birth and the rise and zenith of five solar planets. This indicated that a fetus is aware of movement of a particular planet, and times its birth to be when the planet is rising or at its zenith. You could say the obvious, by saying the child was only following in his parents' footsteps, but then why did it tie its birth time to a particular planet? Science can be amazing, but it's not often almost unbelievable, as in this case.

As I have previously said, Gauquelin was an astrologer who had realised that the whole system of horoscopes was not fit for purpose, saying about printed horoscopes; *Whoever claims to predict the future by consulting the stars is surely confusing himself or someone else*[10.6]. It was his determination to discover the factual scientific evidence about astrology that led him to discover the almost unbelievable truth, that humans inherit an association with one of the five major planets showing it affects their birth time and profession in life. Gauquelin was unable to explain why or how, just accepting it was the reliable part of astrology.

The results of this match between birth times and successful telepathy, led to my realisation that planets emit a quantum carrier of information which solves Gauquelin's problem of how and why a foetus (child) knows when its planet is about to rise. Others have noticed these same similarities in timing, even though they have not seen the far-reaching implications.

So here we have evidence, for the quantum source of the carrier of life's information as well as psi, from two totally different reports, both of which were produced to acceptable scientific standards of evidence. The carrier fills a very important and long-felt requirement for a better explanation of how DNA controls life, and how inheritance functions, and why we experience psi effects.

Microtubules, which are at the heart of quantum biology, are also our sensing elements of sight, sound, smell, and taste, and they are the nerves that transport the information about the body and brain. It has been found that administering a gas, chloroform for example, blocks microtubules, but the mechanism remains unknown. The anaesthetist Stuart Hameroff has pioneered the idea that anaesthetics work by certain gases blocking the action of microtubules, and that a tubulin-binding drug, which is used in modern cancer treatments, works in the same way, and stops or slows cancer growth, working in the same manner as the similar drug Thalidomide.

I have previously hypothesised that the likely carrier of telepathy, instinct and growth information is spinning quantum quartz particles, which raises the question how and why the planets emit these particles. Is it due to the action of gravity on inertia?

Modern physics shows that matter and particles are only waves[10.7], and that gravity is caused by the additional density of the wave medium around an attracting body.[10.8] The effect of this on the earth is that as it revolves around the sun, the nearest part to the sun is subject to drag, making the earth spin once every twenty-four hours. Every planet, like earth, contains quartz. At a quantum level, each particle or atom in a planet is subject to spin due to the action of gravity on inertia, so they become part of a constant shower of spinning quartz.

It's these spinning particles, carrying the DNA identification code, that transmit biological data from parents to the fetus where microtubules act as transmitter and receiver of data. An important observation is that quartz is compatible with DNA, as its crystals are double helixes.

I'm very aware that all this new science can be difficult to follow; perhaps an analogy would help. We understand our mobile phones, where one phone sends a message via the aerial to another one over the airwaves. All phones are listening constantly to a sea of messages. If a call is identified as carrying the receiver's own phone number, then the phone will ring so that the message can be received. And in a similar way, all life can send a message to another life, via primary cilium microtubules, over the biological carrier; and all life is listening constantly. If a message is identified by a fetus as carrying the receiver's own DNA, or a similar DNA, the brain will route the message to the correct stem cell, which will carry out the instructions to correctly differentiate the cell in question. In the same way, messages are sent from one brain to another to build the brain control system, the immune system, and the instinct system, and as I have shown, that mechanism also shares unconscious thoughts known as telepathy. Quantum biology has now achieved the lofty status of being called a science[10.9] but as far as I can see it asks all the questions, while giving no answers.

I am convinced by the probability that a quantum carrier from planets is the carrier of life, and my idea is fully justified by the way the hypothesis fits the evidence.

As I have said many times before, biology and genetics are in disarray and many things are little understood. This situation is like trying to understand a motor car engine with no knowledge of how it works, no knowledge of electricity or petrol, but it has a drivers' handbook written in an unintelligible language. The biologists and a mechanic have taken everything apart and looked at all the pieces which are completely confusing, although there have been many suggestions as to how it might work and how it was assembled. Faced with this problem, it is suggested that the handbook, which has DNA written on the front, is the book of instructions to build the car which has already been assembled by the car dealer. But biologists are unaware of the factory where cars are built. Some biologists are saying the handbook cannot be the building instructions, but still nobody believes them. How cars work, are assembled and designed, is a complete mystery, but why worry, they work well; provided they are kept fed with special fuel. Sorry to say this is where we are with the science of life. We don't understand how life is built, designed or evolved, and why DNA is only the species and family code. This is the truth of the matter, because the science of life and biology is fundamentally flawed, and frankly deficient.

The Paranormal

There is a general misconception with the terminology for the paranormal, although many years ago, the term spiritualism was used to cover anything strange. Today we have; telepathy, clairvoyance, psychic states, and ESP etc; while scientists, use terms like; Remote Viewing (RV), Anomalous Cognition (AC), and Informational Psi (IP). It seems to me psychic states or psi, all use the same mechanism, so let's use the word psi. What is it that's common to all psi states?

First and foremost, it's trance. This is a sort of daydreaming state where consciousness is kept busy with emotion, fear or excitement and the unconscious takes over, almost impossible to describe or achieve, yet a few people are born with it or learn to use it. To people like you and I, it's only another mystery, even though I have had it happen to me a few times in my life, it's still strange and puzzling, making me determined to look for answers.

Science has shown us that deep trance can be measured by the sweat on our hands or electrical pulses at the sides of our head. Max Cade called deep trance the 'awakened mind' or 'state-five' or mystical experience. Albert Einstein had experienced it, noting that; - *'The most beautiful and profound emotion we can experience is the sensation of the mystical. He to whom this emotion is a stranger, who can no longer wonder and stand rapt in awe, is as good as dead. To know that what is impenetrable to us really exists, manifesting itself as the highest wisdom and the most radiant beauty, which our dull faculties can comprehend only in their primitive forms – this knowledge, this feeling, is at the centre of true religion.* He obviously had difficulty writing about his thoughts on this inexplicable experience which results from the deepest trance, and appears to give new insights.

Looking at our everyday state of mind, we are told we tend to slip in and out of trance. If something makes us angry the unconscious instinctive mind takes over. We then do things instinctively that we might not normally do, and regret what was done or said. However, that same light trance can trigger psi, so we might get an emotional message from a friend in trouble, or it could trigger the sense of being stared at.

What could be the mechanism for the sense of being stared at? We are generally aware of our own space and of our looks, personality, and movements on a subconscious level, as they are an important part of social interaction, so we can be aware of someone else watching us, by psi. Tests, relating to the sense of being stared at, revealed astronomical odds against chance.[10.10] If someone is being watched intently, the picture of that person is impressed on the mind of the watcher, and I am sure that the person being watched receives that same impression, by psi. That telepathic message of self creates the sense of being stared at.

Then there are emotional thoughts that can become stored in stone or granite, and subsequently released to be picked up by a person nearby, if they are in trance. This is a haunting or ghost, which can also be self-perpetuating by the receiver becoming emotionally involved, subsequently retransmitting the emotional experience back into the stone. A well-known example of this is the Loch Ness Monster, which is a supposed prehistoric large snake, or monster, living in a Scottish lake. The lake used to be a glacier flowing through granite, which might explain the haunting. Visitors arrive there feeling tired and disorientated by their journey, only to then pick up the emotional view of a monster in the water. My mother did exactly that, and when she pointed it out to my stepfather, he responded, *'Don't be so bloody daft.'*

The evidence of precognition is sometimes linked to psychic effects, where things that are seen in a dream sometimes happen a few days later. My view of this, is that a dreamer sees someone else's plans by psi, before the plans are carried out, and then thinking it to have been precognition. Another example of precognition is when a person's worry about a possible disaster is transmitted to others during a dream, and they then visualise the disaster before it happens. This happened on the Titanic when many passengers experienced emotional dreams before it sank. Many sent letters and telegrams about their forewarnings even before the ship set sail, and one notable passenger, Major Archibald Butt, wrote his will and closed his affairs before embarking on his doomed final journey.

Another example of light trance and psi is called reincarnation, which I believe is the wrong term. In everyday life an emotional event, perhaps an accident, getting lost or trapped, is sent by psi to a family member in trance. In a similar way, in reincarnation where DNAs are sufficiently alike, the fetus may receive a deep emotional message and make an association or link with the person having that similar DNA. This is the same mechanism as inheritance and is why it is reported that fifty per cent of reincarnation cases involve a violent death or murder, so it is the emotion, triggering a psi link with a fetus, sometimes causing birthmarks copying the death marks, as well as a psi link or copying association that continues for four or five years. In fact, 'reincarnation' is no more than the inheritance system wrongly identifying DNA and then connecting with the wrong person. This occurs when there is similar DNA, allowing a telepathic association resulting in physical and mental copying, even to knowledge of homes and people. There is little evidence for this in the West because population movement results in a wide range of DNAs, but in Asia populations have been more static, resulting in less variation in their DNA, and therefore more incidences of 'reincarnation'. This happens when DNAs are similar which opens a psi link between the fetus and a stranger; demonstrating that DNA is a key rather than a book of instructions which I have said many times already; but proving the new core theory is correct.

Now let's look at trance being induced deliberately, in order that the psi mechanism is experienced. It happens in many ways which most of us know about. Examples are, spiritual healing, mediumship and spiritualism, and tests involving card guessing or telephone calling, or picture guessing using the Ganzfeld trance technique.

The U.S. government were concerned about the possible use of psi by foreign countries for information gathering espionage. They set up the Stargate programme, spending about twenty million dollars in 15 years evaluating deliberate psi, calling it remote viewing and informational psi, but then decided; '*There is no evidence that the paranormal would prove useful in intelligence gathering*'. Nevertheless, it did produce plenty of evidence for psi.

The study of healing has produced some quite startling results. Of course, healing doesn't always work, it is a bit hit and miss, although the Bible gives accounts of Christ's healing, so spiritual healing has been practiced for millennia. I attended a healing class for 10 weeks, which was not spiritual, but scientific, when trance was induced in class members, and monitored, but my healing ability was negligible, even though I did have a peak experience.

Later, I interviewed the Chairman of The Confederation of Healing Organisations in the UK, speaking to Bob Charman for half an hour, which you can hear on my podcast number 21. He spoke of healers increasing the germination of seeds, also healers being employed in British hospitals. Bob spoke of the effect of psi in the healing process in which the special state-five brainwave pattern of a healer is copied by the person being healed (healee). He noted their state-five brainwave patterns were identical on each side of both of their heads, what he called 'psi mediated brain synchrony'; Max Cade called it the attuned bilateral fifth-state pattern. What happens is that the healer unconsciously learns of the healee's problems by psi, enabling him to unconsciously send psychic healing information during the time when their brainwave patterns are in harmony.

He noted that Max Cade organised a public demonstration of 400 people, in which the brain waves of healer and healee were seen on instrumentation, coming into bilateral synchrony within 15 minutes.[10.11] This was an important, spectacular presentation, confirming the power of healing by demonstrating the way it works. The only thing that was lacking at the time was an understanding of the mechanism of psi. Bob mentioned a test in Italy, where 12 good meditators all had their brainwaves monitored during meditation. The result was that after a while the brains of 11 were locked in synchrony; more evidence of the amazing mechanism of psi.

Belief in the paranormal has always been mysterious, partly because how and why have been unknown, and partly because psi manifests in so many different ways, as noted above. My new core theory of biology produces the answers as to how and why psi occasionally happens, and in so many different guises; all using the same mechanism of quantum biology. The paranormal is extraordinary, but now understandable, which supports my hypothesis and new core theory. The sceptics say *It can't happen, so it doesn't happen.* Nevertheless, I have shown how, *it can happen, so it does happen.*

Evolution

Despite the writings of Charles Darwin, evolution has remained an enigma, because no explanation has been put forward for inheritance. My new core theory shows how inheritance alters each subsequent generation, according to the needs and desires of parents in a changing environment, allowing for diversity in the members of a species, which leads to its gradual evolution.

Each new member of any species is unique, having unique genes. Where that uniqueness is selected, either by the desires of man or naturally by the desires of the parents, that selection will alter the variety, or even a species, over time. We see this in our own selective breeding of crops, or breeds of dogs from the original wolf. Similarly, female birds select males with bright colours or enormous tail feathers. Examples include birds of paradise or peacocks, but it's the same with a robin or a magpie.

It is only where the selection is continuous that a variety or species will become established, but a single new trait or mutation will breed out to three per cent in five generations and will disappear completely after selection is stopped. For this reason, the ninety-year-old idea, called the Modern Synthesis, which relied entirely on mutations, has been discredited at last,[10.12] leaving a void in the understanding of evolution. That void is only filled by the knowledge that all life is the result of the **direct** copying of parents.

Charles Darwin was unable to explain how the desires of parents were **transmitted** to the next generation, which was the same problem his grandfather raised. The discovery of DNA has only made this situation worse, as there is no acceptable mechanism by which DNA could change. It is only by accepting that life proceeds by direct copying of the previous generation that we have a mechanism (where DNA plays only a selective role) and we can see inheritance, evolution, and diversity in a different light. It is the changes in parental lifestyle and environment, leading to needs and desires being '**transmitted**' to their children, that will cause physical or mental changes in the next generation, known as diversity.

When looking at the beauty of life, many rightly claim that there must be guided evolution and intelligent design. The modern version of this suggests that goal-directed purpose[10.13] is the result of needs and desire, so the members of a species guide their own evolution by providing intelligent design. Darwin's theory initially produced much outrage; I suspect this suggestion might do the same.

Fossil records reveal that sudden, unaccountable changes in evolution have been observed in some species in the distant past, which cannot be attributed to normal slow changes. This is explained by another mechanism for evolution – hybridisation, or natural inbreeding between species, where a cross sometimes leads to a new species (providing it can survive in its environment). So, Darwin only explained one third of the theory of evolution. He had no mechanism for diversity, and he did not understand the implications of hybridisation.

Now that we have a mechanism for inheritance and diversity, the mechanism of evolutionary change in bodily shape and size becomes clear. Not only that, the speed of change is fast. For example, we have changed the wolf into a small dog; a Chihuahua. In the natural world, change due to the environment, predators or food availability is slower, but it is understandable how a leaf-eating deer could change into a giraffe in perhaps four hundred generations, due to its desire for a longer neck, which may be very fast in evolutionary terms. It might total two metres in four hundred generations, or only five millimetres per generation. Some squirrels have learnt to glide by growing webs between their legs. Similarly, some small mammals, such as bats, have learned to fly; some fish take to the air to avoid predators, and, of course, some reptiles turned their fur into feathers.

These changes came about in the natural world over thousands of years, but we have brought about change much faster by selecting crops and farm animals, even managing to alter the temperament of a fox in a decade. This demonstrates that all evolution must be due to the inheritance of traits, needs or desires from parents and relatives, and it's only by means of a new core theory that we can see a mechanism for this inheritance. Looking at the whole of life in this way makes sense of how life evolved. Similarly, looking at science in a new way makes sense of genetics, the brain, instinct, telepathy, mediumship, reincarnation and more.

Resolution by Revolution?

While looking for a vaccine after a pandemic, Griffiths discovered that by moving DNA from one bacterium to another transferred a virus over to another bacterium. This principle was confirmed by microbiologists in the U.S. fifteen years later, then written up as the key to life by the Nobel physicist, Schrodinger.

They were all wrong. It was not the transfer of DNA material but the transfer of DNA barcodes that allowed life to be transferred, which I have recently been able to prove. The ramifications of their error are enormous; biologists have been trying to discover how the instructions for life are written into a code for the last 77 years. They have tried to find out how the instructions could be in the DNA and failed, leading to the European Commission trying to 'rethink the philosophy of biology'. They failed! It's the unique identifying barcode in every living cell that is the key to growth, but it is not the instructions to build life.

Throughout this book I have tried to show the way this could be. Being a lone voice, among the miss-belief of preconceived opinion, has been my challenge, which has been both fascinating, rewarding, and exciting, as each discovery led on to another. The reason for this is that there have been so many anomalies in life that scientists have accepted them as 'not fully understood' to be solved in the course of time and further scientific investigations, which has not happened. The main problem, using a scientific term, is that of paradigm paralysis which is the inability or refusal to see beyond the current models of thinking. What is lacking is an investigation of trans disciplinarity systems theory, applied to biology, instinct and telepathy.

Looking at my journey of discovery, the first was that link between DNA and telepathy, resulting from the realisation that instinct can only be transferred by psi, giving a mechanism for the understanding of telepathy and indeed the whole of life. This 'new core theory' highlighted the fact that inheritance had a new meaning. Similarly, the science of evolution, which has had a deficiency since 1796 is solved by an understanding of the inheritance of acquired and desired characteristics, rather than the bad idea of mutations.

The results of an analysis of 2482 successful psi tests, combined with the analysis of 36,000 birth times, led me to realise that the carrier of psi comes from planets, together with the way in which particles carried life's information between cells by means of their microtubules.

That trail of Making Sense of Life seems almost endless but still has more to explore. Consciousness has been a mystery, but now it can be seen that there is only one brain and mind, repeated in each new generation. Every new mind has the same instinct and mental control system, transmitted across from mind to mind. The evidence of instinct and telepathy shows that the brain is listening all the time for signals from other people, or animals. The resolution of so many anomalies represents a giant leap forward, which means that the science of life will never be the same again.

The *Structure of Scientific Revolutions* was published by the American philosopher of science, Thomas Kuhn, in 1963, when he challenged the nature of scientific progress, saying that day-to-day science misses transformative ideas. He set out how anomalies are ignored until there is a situation of crisis. At this point, a new paradigm may emerge, causing a change in scientific thinking and replacing the previous theory. He said that theory testing and change may only occur when a persistent failure to resolve a noteworthy puzzle has given rise to a crisis. I ask: is my new core theory likely to cause a paradigm change?

APPENDIX

Listen to the author's radio broadcasts entitled Paranormal Is Normal, made for Webtalkradio.

Currently on my website;
https://makingsenseoflife.uk   Podcasts

Or; https://alabone.podbean.com

1   Paranormal is Normal – Science is deluded regarding the Paranormal with Rupert Sheldrake about his book *The Science Delusion*.

2   Paranormal is Normal – Ghosts in the city are Normal, with a guide to the ghost tours of Prague.

3   Paranormal is Normal – Some Astrology cannot be Denied. How can Gauquelin's work on astrology relate to the paranormal?

4   Paranormal is Normal – A transplant heart carrier's lifestyle characteristics. The evidence of the occasional influence of heart transplants.

5   Paranormal is Normal – Quantum Science and Angels that Heal, with the editor of *Quantum Angel Healing* and her experiences of healing.

6   Paranormal is Normal – Stage Hypnosis Shows are great entertainment, with a member of the Federation of the Ethical Stage Hypnotists.

7   Paranormal is Normal – Praying is Normal; even Praying in Tongues, with a church minister about speaking in tongues and healing.

8   Paranormal is Normal – The Art of Dowsing is like Magic - but quite Normal, with a retired president of the American Society of Dowsers concerning his experiences and thoughts on dowsing.

9   Paranormal is Normal – Trance and Mental Illness, with a well-known therapist who has organised the training of 200,000 British therapists.

10 Paranormal is Normal – It's NOT a Mystery. So how can the Paranormal be Understood? My thoughts on how and why the 90 different paranormal experiences are part of life.

11 Paranormal is Normal – Remote Viewing – How is it Done? With Gwyn Hocking who explains how some people can 'see' remotely with the power of the mind.

12 Paranormal is Normal – End of Life Experiences or NDE's, with Peter Fenwick who has studied and written about Near Death Experiences (NDEs).

13 Paranormal is Normal – Traditional Oriental Reiki, with Guido Terzaghi to explain the spiritual and paranormal side of Reiki.

14 Paranormal is Normal – Unconscious Connecting – or Telepathy, with a teacher who instructs autistic children in non-verbal communication.

15 Paranormal is Normal – Talking to a Medium with the gift of Healing and Spiritual Surgery, with a spirit medium talking about spiritual healing.

16 Paranormal is Normal – Training with a Mind Mirror to achieve altered states of consciousness, with Judith Pennington who trains people to use a special EEG.

17 Paranormal is Normal – Past Life Therapy hypnosis is used to regress, to discover how a previous life might affect one's mental and physical health in this life.

18 Paranormal is Normal – Lucid Dreaming:- Awake when Asleep, with a lucid dreamer who has been studied by Stanford University's Division of Sleep Medicine.

19 Paranormal is Normal – A Psychic is able to act in Many Ways, with a psychic who has worked in various areas of the paranormal.

20 Paranormal is Normal – Floatation Tanks –the way to deep meditation discussing how deep trance can be achieved.

21 Paranormal is Normal – Healing is perfectly Normal and we begin to know why, with the Chairman of the Confederation of Healing Organisations to discuss the many aspects of healing.

22 Paranormal is Normal – Helping People learn to Heal Themselves, with the author of *Empowered Healing* to examine the power of hypnotherapy, trance, Reiki and spiritual healing.

23 Paranormal is Normal – Medium of the century: from Brazil. This episode concerns Chico Xavier, who used automatic writing for healing and other psychic powers.

24 Paranormal is Normal – These machines aid Meditation and train Psychic Ability, with the managing director of Meditations Ltd whose machines use light and sound for deep trance training.

25 Paranormal is Normal – The Artistic Sleepwalker – Lee Hadwin. My guest makes amazing

drawings unconsciously when asleep. Some have been sold for charity.

26   Paranormal is Normal – Even a Scientist and Business Executive can be Psychic. Since her retirement, Jude Currivan has travelled the world and written about her own experiences as well as many others she has met.

27   Paranormal is Normal – The many different Forms of Healing. Are they all the same? Phil Young has studied and practiced many different forms of healing; but healing is healing.

28   Paranormal is Normal – The Science, and understanding, of Twin Telepathy. The details revealed.

29   Paranormal is Normal – Can Trance and Telepathy explain ALL ? Are all psychic abilities the result of a natural part of life?

30   Paranormal is Normal – Exploring the power of Healing and Reiki, with Aram Kong who uses acupuncture and speaks of its uses.

31 Paranormal is Normal – It seems that UFOs are everywhere – but what are they really? With an ex-colleague of mine who is heavily involved in the British UFO Research Association

32   Paranormal is Normal – Spiritualism is Normal. But let's look at the detail. With a Minister and Director of the Spiritualist National Union

33   Paranormal is Normal – Clairvoyance is considered impossible but, for a few, it is normal, with a personal friend who has had to live with the amazing effects of clairvoyance.

34   Why do Folks hear music when someone is Dying ? Near Death Experience and music. A study of the emotional effects of death.

35   Paranormal is Normal – A million People Hypnotized from CD's, with a top hypnotist who produces hypnotherapy CDs for self-use.

36   Paranormal is Normal – Yes: but all too often it's faked. My thoughts on the reasons.

37   Paranormal is Normal – Origin of Dreams and trance, with the author who has revealed the science of why we dream.

38   Paranormal is Normal – I Believe in Ghosts, with the author of *Psychic Quest*, an in-depth study of ghosts and hauntings.

39   Paranormal is Normal – Hypnosis really works, with a hypnotist about the how and why of his work.

40   Paranormal is Normal - Animal Telepathy, with Rupert Sheldrake who discusses his book detailing human and animal telepathy.

Throughout the broadcasts I use the word telepathy to mean several things, whereas the words 'brain sharing' would have been more applicable.

Acknowledgements

I have been very aware that this book has relied entirely on the work and ideas of many scientists and thinkers. Although some of the references are noted, most are not, so I gratefully acknowledge that the evidence is the work of others; all I have done is to offer an alternative interpretation of that evidence. Also, I would like to thank my editor Liz Gordon

Richard Alabone

See website;   https://makingsenseoflife.uk

Making Sense of Life

References

Chapter 1 An Introduction
[1.1] Wikipedia *Morphogenetic Fields* 13-02-2020
[1.2] Erwin Schrodinger *What is Life* 1992 pp 21-22
[1.3] Thomas S. Kuhn *The Structure of Scientific Revolutions* 4th Edition 2012

[1.4] Philippe Huneman & Dennis M. Walsh *Challenging the Modern Synthesis: Adaptation, Development, and Inheritance* p 1

Chapter 2 The Things That Don't Make Sense
[2.1] Rupert Sheldrake *New Science of Life: The Hypothesis of Morphic Resonance* 1981 pp 19-21

Chapter 3 The Wonders of Biology
[3.1] Bruce Alberts *Molecular Biology of the Cell* 5th Edition 2002
[3.2] Scott F. Gilbert *Developmental Biology 7th Edition* 2003
[3.3] Ibid p 4
[3.4] *Microtubule* Wikipedia 29-01-2013
[3.5] Tim Spector *Identically Different: Why You Can Change Your Genes* 2012 p 16
[3.6] Nessa Carey *Junk DNA* Icon books 2015
Chapter 4 Brain Sharing: Telepathy
[4.1] Richard Alabone https://makingsenseoflife.uk

podcast 33

4.2 Rupert Sheldrake *Telephone Telepathy with the Nolan Sisters* YouTube nolan sisters telephone telepathy
4.3 Upton Beall Sinclair. *Mental Radio* (Forgotten Books) 2008
4.4 Thomas S. Kuhn *The Structure of Scientific Revolutions* 4th Edition 2012 p144

Chapter 5 It's the Planets
5.1 James P. Spottiswoode *Apparent Association Between Effect Size in Free Response Anomalous Cognition Experiments and Local Sidereal Time* Journal of Scientific Exploration, Volume 11, No. 2, 1997
5.2 Michel Gauquelin *Cosmic Influences on Human Behaviour* 1973
5.3 Nick Kollerstrom *How Ertel Rescued the Gauquelin Effect*. Correlation Volume 23, Issue 1, p 3
5.4 John D. Ralphs *After Gauquelin: What Next?* Network Review, Journal of the Scientific and Medical Network, Issue 99, p 15, 2009

5.5 Weizmann Institute of Science
Biological Molecules Select Their Spin
https://wis-wander.weizmann.ac.il/space-physics/biological-molecules-select-their-spin

Chapter 6 The Philosophy of Consciousness

[6.1] PhilPapers *Science of Consciousness* 21-10-2016
https://philpapers.org/browse/science-of-consciousness

[6.2] James Le Fanu *Why Us?: How Science Rediscovered the Mystery of Ourselves* 2010

[6.3] Richard Alabone *A Squirrel's Tale*
The Essex Beekeeper No. 546 pp 14-16

[6.4] Joseph Griffin *The Origin of Dreams* 1997

[6.5] Richard Alabone_https://makingsenseoflife.uk
podcast_37

Chapter 7 The Spooky List

[7.1] Joe Griffin & Ivan Tyrrell *Hypnosis and Trance States: A New Psychobiological Explanation* 1999

[7.2] Maxwell Cade & Nona Coxhead *The Awakened Mind:* 1979 pp 24-25

[7.3] Ibid p 44

[7.4] Richard Alabone *Sublimity: The Sublime Nature of Man* 2009 pp 71-80

[7.5] Maxwell Cade & Nona Coxhead *The Awakened Mind:* 1979 p 48-49

[7.6] Wikipedia *Rapid Eye Movement Sleep* 24-03-2021

[7.7] Richard Alabone https://makingsenseoflife.uk
podcast 18

[7.8] Richard Alabone *Sublimity* Author House 2009 p58

[7.9] Nona Coxhead *The Relevance of Bliss* 1985 p 2

[7.10] Richard Alabone   https://makingsenseoflife.uk podcast 7

[7.11] Ibid *End of Life Experiences or NDE's* No12
[7.12] Ibid *Even a Scientist Can be Psychic* No 26
[7.13] Ibid *Praying Is Normal; Praying in Tongues* No 7
[7.14] Ibid *Healing Is Perfectly Normal* No 21
[7.15] Ibid *Lucid Dreaming-Awake When Asleep*   No 18
[7.16] Ibid *The Many Different Forms of Healing. Are They All the Same?* No 27

Chapter 8 The More You Look the More You See

[8.1] George N. M. Tyrrell *Apparitions* 1973
[8.2] Ibid pp 18-24
[8.3] Ibid p 33
[8.4] Richard Alabone   https://makingsenseoflife.uk

  Clairvoyance is considered impossible   No33

[8.5] Ibid *I Believe in Ghosts* No38

[8.6] Natalie Osborne-Thomason *Psychic Quest* 2002 p 147
[8.7] Elizabeth Lloyd Mayer *Extraordinary Knowing:* 2008
[8.8] Robert A. Charman M*ediums and a Possible Source of Communication:* Journal Society for Psychical Research, Volume 79, Issue 920, p 186, 2015

[8.9] Larry Dossey *One Mind: How Our Individual Mind is Part of a Greater Consciousness* 2013   p 147

[8.10] Ibid p 136

[8.11] Ibid p 53

[8.12] Ibid p 57

[8.13] William Braud & Marilyn Schlitz *A Methodology for the Objective Study of Transpersonal Imagery* Journal of Scientific Exploration Volume 3, No. 1, pp 43-63, 1989

[8.14] Jeanne Achterberg *The Role of Imagery in Healing: Shamanism and Modern Medicine* 1991

[8.15] William G. Broad *Distant Mental Influence of Rate of Homolysis of Human Red Blood Cells* The Journal of the American Society for Psychical Research Volume 84, No. 1, pp 1-24, 1990

[8.16] William G. Broad, *Experiments with Mathew Manning* The Journal of the American Society for Psychical Research Volume 50, pp 199-223, 1979

[8.17] YouTube *Secret CIA Psychic Lab Experiments with Uri Geller at Stanford University* 2010

[8.18] Ian Stevenson *Where Reincarnation and Biology Intersect* 1997

[8.19] Ibid p 110

[8.20] Ibid p 112

[8.21] Ibid p 115

[8.22] Richard Alabone https://makingsenseoflife.uk No4

Chapter 9 Evolution and Inheritance

[9.1] Alister Hardy *The Living Stream* 1965 pp 55-56

[9.2] Ibid p 81

[9.3] Ibid p 258

[9.4] Ibid p 231

[9.5] Ibid pp 160-161

[9.6] Ibid p 255

[9.7] Ibid pp 257-258

[9.8] Macroevolution.net *Bird x Fish*
www.macroevolution.net/carp-wagtail-hybrids.html

[9.9] Macroevolution.net *The Hybrid Hypothesis.
Human origins: Are we hybrids?*
www.macroevolution.net/human-origins.html

Chapter 10 Discussion

[10.1] Scitable *Isolating Heredity Material: Frederick Griffith, Oswald Avery*

[10.2] Epigenetic Inheritance: Concepts
https://www.ncbi.nlm.nih.gov/

[10.3] Roger Penrose *Shadows of The Mind*
Oxford Press 1994

[10.4] The Miracle of Fetal Development
https://www.hli.org/

[10.5] Michel Gauquelin *The Cosmic Clocks* Chaucer Press p166

[10.6] Ibid p85

[10.7] Milo Wolff *Schroedinger's Universe and the Origin of the Natural Laws* 2008

[10.8] Ibid p 63

[10.9] The Scientist *Quantum Biology* 01/06/2019

[10.10] Rupert Sheldrake *The Sense of Being Stared At* Hutchinson 2003 p170

[10.11] Maxwell Cade & Coxhead *The Awakened Mind:* 1979 p197

[10.12] Philippe Huneman & Denis M. Walsh *Challenging the Modern Synthesis:*

[10.13] Ibid p 251

XXX

Printed in Great Britain
by Amazon